Aristotle's

Metaphysics

Metaphysics

Aristotle's
Metaphysics

a new translation by
Joe Sachs

Green Lion Press

Santa Fe, New Mexico

Manufactured in the United States of America.

Published by Green Lion Press,
1611 Camino Cruz Blanca, Santa Fe, New Mexico 87501 USA.

Telephone (505) 983-3675; FAX (505) 989-9314;
Orders (within USA only) 1-800-852-1373.

mail@greenlion.com
www.greenlion.com.

Green Lion Press books are printed on acid-free paper. Both softbound and clothbound editions have sewn bindings designed to lie flat and allow heavy use by students and researchers. Clothbound editions meet the guidelines for permanence and durability of the Committee on Production Guidelines for Book Longevity of the Council on Library Resources.

Printed by Edwards Brothers, Inc., Ann Arbor, Michigan.

Cover illustration based on portrait of Aristotle in Raphael's "School of Athens." Cover design by Dana Densmore and William H. Donahue with help from Nadine Shea.

Cataloging-in-Publication Data:

Aristotle
Metaphysics / by Aristotle
            translation by Joe Sachs

Includes index, bibliography, introduction, and notes.

ISBN 1-888009-03-9 (sewn softcover binding)
ISBN 1-888009-02-0 (cloth binding with dust jacket)

1. Aristotle, Metaphysics, English. 2. Philosophy. 3. History of philosophy.
4. Classics. 5. History of Science.
I. Aristotle (384–322 B.C.E.). II. Sachs, Joe. III. Title.

B434.A5 S23 1999

Library of Congress Catalog Card Number 98-83157

Contents

The Green Lion's Preface

The Green Lion Press is delighted and honored to be presenting Joe Sachs's translation of Aristotle's *Metaphysics*. This book follows up the success of Sachs's translation of Aristotle's *Physics*, published by Rutgers University Press, and addresses the demand which that translation has generated for Sachs translations of more of Aristotle's central works.

Sachs's translations bring distinguished new light onto Aristotle's works. Joe Sachs translates Aristotle with an authenticity that was lost when Aristotle was translated into Latin and abstract Latin words came to stand for concepts Aristotle expressed with phrases in everyday Greek language. When the works began being translated into English, those abstract Latin words or their cognates were used, thus suggesting a level of jargon and abstraction—and in some cases misleading interpretation—that was not Aristotle's language or style.

Joe Sachs's introduction to this book tells the history of the translation of Aristotle in general and *Metaphysics* in particular. It offers insight into and clarity about this work called *Metaphysics*, perhaps more properly titled *First Philosophy*—this crown of Aristotle's thinking. Sachs shows how by the use of dialectic Aristotle is able to ignite our imaginations to bring us to a greater understanding of cause and being than would be possible by following a dry syllogistic line. In his introduction Sachs traces Aristotle's development of his project through the whole of *Metaphysics* showing us the integrity and wholeness of this work, which more superficial and less sensitive thinkers have sometimes accused of being fragmented and inconsistent. As additional aids to grasping the unfolding of the argument, Sachs offers his own titles for each Book and provides a separate annotated outline of the whole argument. These introductory essays are not only greatly helpful to the reader but make an important contribution to Aristotelian scholarship.

In addition to offering a groundbreaking new translation and these introductory materials, we have designed this book to be easy to read, study, teach with, and use as a basis for discussion. Having ourselves read, studied, taught, and discussed this text with students and colleagues over many years, we know what we as users needed and wanted in presentation and layout, and we have taken pains to provide those things.

As readers we wanted a sewn binding so that pages would not fall out even on repeated readings. We wanted high quality paper and generous margins for making notes. We wanted a sturdy cover that would not curl or fray when the book was carted around. We wanted a glossary that thoroughly explained translation choices. We wanted footnotes, not end notes, so that we could see at a glance what the note said and easily either read it or defer attending to it without breaking the flow of our reading of Aristotle. We wanted the footnote number in the text to be large or bold enough that it could be found if later we came back to a footnote and wanted to find the place in the text where the footnote was invoked. We wanted Bekker page number and line number ranges, when cited, to be specified in full so as not to be ambiguous or misleading. We wanted a typeface and choice of leading (spacing between the lines) that made the reading easy so that the hard thinking could be directed to what Aristotle was inviting us to address, not to the reading process. We wanted an index that was thoughtfully prepared by someone who had thought deeply about the text and the issues it raised.

We wanted footnotes citing other primary source texts to make it possible for the reader to find the quotation or reference in an edition he or she might have in hand or be able to find in the bookstore or college library. One might appear to have met the responsibility of citation by a reference to page numbers in an edition out of print and no longer available, or in an obscure old edition perhaps venerable and of antiquarian interest and charm but now difficult or impossible to locate. But in fact such a reference may not serve the reader or allow an independent check on the interpretation being made by the author. We wanted citations to the relevant book, chapter, section, paragraph or line numbers within the primary source text, so that the quotation or text being paraphrased could found in whatever edition or translation we owned or could conveniently consult.

As scholars, we wanted all those features and other things as well. We wanted the Bekker numbers[*] and the Book and Chapter numbers

---

[*] Page and column numbers of the Bekker edition and line numbers of the Oxford edition of Aristotle's works. See Part IV of the Introduction for explanation of these.

easy to find and follow. We wanted a glossary of the Greek terms, in Greek alphabetical order, as well as a glossary of the English terms. We wanted to have both the Roman numeral book numbers and the Greek letter Book numbers on the Book titles and running heads since the Books are sometimes referred to by one and sometimes by the other nomenclature, and the two Books entitled *alpha* (A and α) make conversion confusing even for those who remember the Greek alphabet.

As teachers and participants in discussions of the text, engaged with others using other translations, we wanted all that and in addition anything and everything to let us quickly and accurately find the place in the text to which someone was referring. We wanted to be able to use the book in a seminar or lecture question period and be able to find a place mentioned easily enough that we didn't have to take our attention away from the discussion and put it into the hunt.

For the occasions when an interlocutor made reference to a bit of text by Bekker page and line numbers, we wanted those numbers in the margins, not embedded so discreetly in the text that repeated scanning of the page failed to find them. We wanted every one of those marginal references spelled out in full, repeating the citation of the page number with every line number, so that we didn't have to turn pages back or ahead to find out on what page the orphan line numbers we were looking at appeared.

For the occasions when an interlocutor made reference to a bit of text by Book and Chapter numbers, we wanted both in the running heads on every page (with both Roman numeral and Greek letter book numbers) so that it could be turned to easily—without having to do as we have done and take each new translation or edition of Aristotle and go through writing those things in by hand on every page.

All these things are provided in this edition. We have made the book we always wished we had when we read and used Aristotle, and we hope that these design features will please you and serve you as well as they will us and that they will make your engagement with Aristotle direct and without external impediments.

Dana Densmore and William H. Donahue
for the Green Lion Press

As soon as one's purpose is the attainment of the maximum of possible insight into the world as a whole, the metaphysical puzzles become the most urgent ones of all.

—William James

Introduction

I. Ways of Writing and Ways of Being

The Ordering and Content of Aristotle's Inquiry

The word *metaphysics,* when heard by most people, is apt to raise a smile of the sort reserved for innocent souls who are harmlessly deluded. People who have read philosophic books are likely to think of the metaphysical tradition as something obsolete, which virtually every philosopher of the last four hundred years has found some new way to reject. And even those who devote themselves to studies within the metaphysical tradition tend to think of Aristotle's book called the *Metaphysics*[1] as an unfortunate accident, put together without rhyme or reason by some editor who did not realize that its parts are mutually incompatible. These three opinions do not depend on one another, but when the error of the third one is exposed, the previous two may begin to weaken. If Aristotle's *Metaphysics* is coherent, it might also be worthy of serious attention, in our time as much as in any other, and not by scholars only, but by all thoughtful people.

Two mistakes give rise to the widespread opinion that Aristotle's *Metaphysics* is not a whole. One of them is the belief that written treatises must always be conceived deductively, even if they are presented with their highest assumptions given last. The other is

---

[1] The name comes from librarians, and reflects the order of our thinking: *ta meta ta phusika* are the things that come after the study of physics. Aristotle's name for the topic is first philosophy, reflecting the inherent order of things.

the belief that, in the first place, all wholeness of thinking must be logical. The mistake mentioned first comes from a failure to understand dialectical inquiry, while that mentioned second comes from a failure to reflect upon the structure of organic wholes. These particular failings are instances of a more general problem. Often readers use insufficient imagination in coming to terms with a text remote in time, for which the habits of more familiar writers cannot be assumed. The Hebrew Bible, for reasons different from but not wholly unlike those that affect the *Metaphysics*, has also suffered at the hands of its unimaginative historically inclined scholars. A more sensitive reader[2] has recently proposed that the writers of the Bible

> had certain notions of unity rather different from our own, and that the fullness of statement they aspired to achieve as writers in fact led them at times to violate what a later age and culture would be disposed to think of as canons of unity and logical coherence. The biblical text may not be the whole cloth imagined by premodern Judeo-Christian tradition, but the confused textual patchwork that scholarship has often found to displace such earlier views may prove upon further scrutiny to be purposeful pattern.

As with the Hebrew Bible, the various parts of the *Metaphysics* abound in repetitions, overlapping treatments of related topics, gaps between successive passages, and plainly contradictory statements. But while the books of the Bible have been carved up, disassembled, and assigned to various sources, the *Metaphysics* has never been accused of multiple authorship by anyone whose arguments were widely credited. The near certainty of single authorship has led scholars to conjecture that Aristotle himself was at least two different thinkers at different stages of composition, that he "developed." And there is no question that the composition of the *Metaphysics* was not a single act; the work is compiled from a number of separately composed pieces. A modern author who put together many short writings into a longer whole would be likely to remove repetitive or overlapping passages, fill gaps, and either explain or disguise any contradictions, but such an author would not be shaping an extended dialectical inquiry, which is what Aristotle did in shaping the *Metaphysics*.

---

[2] Robert Alter, *The Art of Biblical Narrative*, Basic Books, 1981, p. 133.

The notion of dialectic has been construed in many ways in the secondary literature about Plato and Aristotle, and it is important to be clear about how it is being understood here. Dialectic is characterized in various ways in different Platonic dialogues, and the reason for the differences is inherent in dialectic itself. I take as crucial a passage in the *Meno*, at 75 C–D. There, dialectic is explained as the way of doing things that suits friendly conversation about serious questions. Unlike debate, where the aim is victory in verbal combat, dialectical speech cannot be content to say something true, but must get at the truth only by way of things the other person already understands and acknowledges.

Scholars usually contrast an Aristotelian dialectic to a Platonic one, but there is no need to think of them as formal procedures, or as different from one another. One need only ask how the benefits of the sort of conversation Socrates describes in the *Meno* can be attained in a written work in which no particular pair of people is feigned as conversing. Aristotle answers this question in the *Topics*, at 100a 30–100b 23. One must reason only from "things that seem true to everyone, or to most people, or else to the wise, and of the latter either to all of them or most of them or to those who are best known and most respected." By writing in this way, or reading things written in this way, one will not only gain agility in thinking and become better at conversation, but one can also get at the heart of all knowledge, since dialectic "contains the road to the starting points of all pursuits." (101a 25–101b 4) Dialectical reasoning does not set down permanent beginnings such as, for example, David Hume's declaration that all knowledge must derive from sense impressions (in his 1739 work *A Treatise of Human Nature*, Bk. I, Part I, Sec. I). A dialectical inquiry might assume some opinion that equates knowledge with perception (which is just what happens in the first half of Plato's *Theaetetus*), but it would do so in order to try it out and test it.[3] This is the humble meaning of that passage in Plato's *Republic* (511 B) in which Socrates assigns dialectic to the fourth and highest part of his divided line, and to those knowable things "which speech itself gets hold of by means of its power of conversing, making its suppositions not ruling beginnings

---

[3] Dialectic is *peirastikē*—tentative (*Metaphysics* 1004b 25)—and *exetastikē*—probative (*Topics* 101b 3).

but in fact supports, like scaffoldings and springboards, in order to go up to what is beyond supposition at the beginning of everything."

In the *Nicomachean Ethics,* in a passage commonly misunderstood as a rejection of the "Platonic forms," Aristotle praises Plato for inquiring whether the philosophic road is down from or up to first principles (1095a 32–34). This distinction alone should make one wary of using the phrase "philosophic system" indiscriminately for the work of any thinker whose reasoning is coherent and comprehensive, and especially for that of Plato and Aristotle. In his 1637 *Discourse on Method* (near the end of Part II), René Descartes proposes modeling all knowledge on the long chains of reasoning used by geometers, claiming that this project of a system of philosophy had never been undertaken before.[4] It was certainly not undertaken by Aristotle, even though he worked hard to learn everything that is knowable by the effort of thinking. All Aristotle's writings are connected, and hence in the root sense of the word "system" they do stand together, but they do not compose what is ordinarily meant by a system of thought. The *Metaphysics* stands higher in the order of knowledge than does the *Physics,* say, but the latter stands nearer to ordinary experience and earlier in the order of inquiry. That fact alone introduces a deep ambiguity into what it means for either of those inquiries to come first, but that in itself would not prevent them from being deductively ordered, even if dialectically discovered. But the ways in which each inquiry depends upon the other are not so neatly distinct in Aristotle's practice.

Aristotle's writings always undertake the work of the learner, for whom the preliminaries have not been fully settled and the consequences are not yet fully determined. The totality of such inquiries is not like a map from which coordinates can be taken, but more like a living community, in which all the interactions of part and whole are mutual. Some readers of Aristotle believe that he taught the "philosophic sciences," in accordance with a model he described in the first book of the *Posterior Analytics.* Such a reading mistakes a source of life for a rigid, completed structure. What Aristotle discusses in the first book of the *Posterior Analytics* is demonstrable knowledge. The very

---

[4] *The Philosophical Writings of Descartes,* Cambridge University Press, 1985, Vol. I, pp. 120–121.

fact that some knowledge can be attained by demonstration makes it clear, according to Aristotle, that not all knowledge is demonstrable, and that the nondemonstrable sort must take precedence, since it must be present before demonstration can begin (72b 19–25). If all knowledge is connected, what becomes most needful for us is the upward way of reasoning (cf. 82a 23–24) that travels from where we are to the things that can transform our opinions and perceptions into knowledge. The second book of the *Posterior Analytics* is devoted to this role of demonstrative reasoning not to deduce consequences of something already assumed, but to find the middle terms that connect apparent facts to the causes that make them understandable (90a 34–35). The things that we seek to know cannot be demonstrated, but can become clear to us only if we work out the demonstrations they make possible (93b 15–28). The *Posterior Analytics* is itself a dialectical inquiry, which explores the nature of demonstration not in order to imitate or prescribe it, but to make use of it within the quest for knowledge.

But even if the motion of the *Metaphysics* is upward, there seems no reason for it to begin that motion over again more than once. But the *Metaphysics* plainly has more than one beginning. Its first two books are both numbered one (with upper and lower case alphas), so that the remaining books all bear letters normally indicating a number that is one less than their own place in the sequence. But in content, the *Metaphysics* begins over again much more than twice. Of its fourteen books, only Books VIII, IX, and XIV are *not* new beginnings. The eleven sections of the whole inquiry are not set end-to-end like bricks in a row, but are woven together like threads in a complex design. The weave is a tight one and the cloth is whole, not for reasons of formal construction but because the contents of the parts work together like the organs of a living body, which are wholes subordinated to one encompassing whole.

The dialectical purpose within each section is not difficult to find. Book I goes through the things previous thinkers have said about "the causes that originate things"; even though Aristotle has worked out in his *Physics* an account of four kinds of cause, such a review will either lead to a revision of that account, or establish it more firmly (983a 24–983b 6). Book II is an independent reflection on the inquiry into truth, linking truth with being (993b 27–31), anticipating the finitude of both, and warning against the demand for either too much or too little precision. Book III is an extensive collection of preliminary impasses about the knowledge of being, since "those who inquire without first coming to an impasse are like people who are ignorant of which way they need to walk" (995a 34–37). The organization of the whole work will be sketched out following this introduction, but for now one more example of the dialectical character of each separate section will suffice. Books VII through IX form one continuous examination of what being is for perceptible things. At the end of Book VII, Chapter 3, Aristotle explains that these are not the only beings there are, and not the most knowable beings, but only the most familiar and the easiest to agree about. His intense study of them is made for the sake of passing beyond them, to what is inherently more knowable, but less well known to us (1029a 33–35, 1029b 3–12).

It is already abundantly clear that the dialectical ascent of the *Metaphysics* is not simply a deduction in reverse. Various roads are traveled, that are partly parallel, partly divergent, but always finally convergent; the goal is not simply to get to an end but to get there well, to cast a variety of lights on the way that it is an end, to reinforce previous conclusions with related observations, and to reflect the true complexity of the topic, in which there is no reason to expect neatness. Such a journey involves repeatedly stopping, backing up, and partially retracing some ground that has already been covered in a different way. Some pairs of contradictory formulations people find are within single sections of the inquiry, and are merely instances of the fact that dialectic begins with opinions that seem true, and then revises them. For instance, in Book VII, a certain kind of being is said first to include the parts of animals and plants and also such homogenous bodies as fire, water, and earth (1028b 9–15), and later to exclude those same things (1040b 5–10). In context the later statement does not contradict the first any more than afternoon contradicts morning. But there are also contradictions that people think they see between things in different sections of the inquiry.

The largest stumbling block among the latter sort of apparent contradictions is Aristotle's varying formulation of the topic of the inquiry. What is the *Metaphysics* about? Among several answers to this question, two recur and stand out. The inquiry aims at understanding "what is, insofar as it is" or "being as being" (*on hē on*, 1003a 21–22), but it also aims at understanding "the sources and causes of being" which must be "things that are both separate and motionless," and therefore "divine" (1025b 3, 1026a 15–18). The first formulation begins Book IV, and the second begins Book VI. In virtue of the second characterization, the whole study is called by Aristotle "theology" (1026a 19); on account of the first, it is called "ontology," but only by others, and much later. It might seem strange that anyone ever considered being itself, as being, to be the same topic of study as an unchanging supreme being, but ancient scholars, for centuries after Aristotle wrote, believed just that. But does it not seem, by mere logic, that what belongs to everything must be emptiest in content? Hegel takes this thought one step further, concluding "The true state of the case is[:] . . . Being, as Being, is nothing fixed or ultimate . . . But this mere Being, as it is mere abstraction, is therefore the absolutely negative . . . is just *Nothing*."[5]

Hegel's understanding of being as being is the logical extreme of the idea that being as such cannot be the particular way of being of anything, but must be common to everything. Before surrendering to any such interpretation, however, which forces us to think that Aristotle has thrown together the emptiest and fullest meanings of being in an incoherent way, one should notice that, when the expression "being as being" is introduced at the beginning of Book IV, it is immediately expanded with the phrase "and what belongs to this in its own right." Aristotle often links with an "and" two formulations which refer to the same thing, determine that thing jointly, and limit the meanings of each other. The pairing of "separate and motionless" mentioned in the last paragraph functions in that way, and one may find countless other examples almost anywhere in Aristotle's writings. I suggest reading only the first "being" as the antecedent of "this," so that it is modified by two phrases that function as synonyms in the context: being *as being* is being *as it is in its own right.* Everything that we can perceive or even think *is* in some way. The perceptible things *exist,* that is, have being in a place and at a time, but anything thinkable can also be present in some way and have effects, and must also have its way of being. But cutting across that distinction among beings is another distinction between what is dependent and what is independent, or what is derivative and what stands on its own. A leaf has being, but in a way that depends on the being of a plant. The whole plant, in relation to the leaf, can more so be said to have being in its own right. This is not just

---

[5] G. W. F. Hegel, *Logic* (from his *Encyclopedia*) Sections 86–87. (*Hegel's Logic*, Oxford University Press, 1975, pp. 126–127.)

a verbal or logical distinction, but recognizes the fact that the plant actively keeps itself in being, by seeking out and soaking up water and sunlight and soil minerals. The plant's relation to its own being is active and self-maintaining, at least to some extent, and to that extent it displays being as such, or being as being.

That this is what Aristotle means by the phrase "being as being" is confirmed by abundant detailed evidence (though scholars can always quibble about that) as well as by the coherence this reading gives to the whole of the *Metaphysics*. Two other sorts of corroboration are worth mentioning. First, there is the fact that the other way of reading the phrase, as that which belongs to anything whatever that can be said to be, constitutes a single class of all beings which Aristotle explicitly says is an impossibility. Twice (at 998b 22–27 and again at 1059b 32–36) he shows that logical impossibilities follow from trying to formulate a general class that includes everything. That is why he argues that being is meant in irreducibly many ways (the eight so-called categories), and the question of which way of being is primary, not in a logical or linguistic sense but causally, gives the main impetus to his central argument.[6] Second, the phrase *on hē on* echoes a phrase used frequently in the Platonic dialogues, *ontōs on,* for that which most truly is. It means being in the manner of being, beingly being, much as we sometimes speak of an actor's actor or the cream of the cream. This doubling is a natural way to intensify whatever is meant, as it unquestionably does in the Platonic phrase audible in the background of being as being, even though its effect might at times, as in "human, all too human," also be a dilution.

An analogy might clarify what is at issue. Herodotus, in the *Persian Wars,* describes the battle of Thermopylae, in which three hundred Spartan soldiers, with a handful of allies, confronted hundreds of thousands of Xerxes's troops. In Book VII, Chapter 210, Herodotus writes that Xerxes, watching the fighting, realized that he had many *anthropoi* but few *andres.* The former word refers to mere human beings, the latter to those with manly excellence. We could use the phrase "human beings as such" to mean the former, who display the lowest human tendencies, but we might also use it to mean the latter, the ones

---

[6] The primary meaning of being, or of any fruitfully ambiguous word, unifies the rest of its meanings, but in the manner of a cause in relation to effects of different kinds, not in the manner of a common element unequivocally present in each.

who display in full activity the highest human capacities. These highest capacities may, of course, not be military or masculine. In Homer's *Iliad*, for example, Achilles fails to achieve the utmost in military glory, but then rises higher in Book XXIV by recognizing and honoring the dignity of his humiliated enemy. The *Iliad* is not an inquiry, but it is like one in that Achilles is in quest of something throughout it. That quest is not dialectical, but it resembles dialectic in that the conception of the thing sought is first a borrowed and conventional one, but is then transformed by what is learned in the pursuit of it. Finally, what Achilles discovers is not a causal structure, but like such a structure it also is a vision of what is highest in its kind, that illuminates everything in that kind in common.[7]

The highest being, the cause of being, and being itself, in its own right as being, all fuse into a single object of inquiry because, according to Aristotle, a thing can only be understood when one knows the cause that makes it what it is. Whatever the derivative thing might be will be present more eminently in its cause and source. He writes in Book II (993b 22–26), "without the cause we do not know the truth, but each thing is what it is most of all, and more so than other things, if as a result of it the same name also belongs to those other things. (For example, fire is the hottest thing, since it is also responsible for the hotness of the other things.)" The thinking that grasps the unity of Aristotle's endeavor is already philosophic thinking, and in getting hold of what the *Metaphysics* is about, we have begun taking part in the inquiry along with its author.

But if we are engaged in this inquiry, what exactly are we doing? The Biblical tradition, like the metaphysical one, is also concerned with uncovering the highest being and source of all other beings, but there that uncovering is understood as the preservation of the stories of men and women to whom that being chose to reveal itself. And, for the sake of argument, we might assume that for some people the highest being and cause of all other beings is the atom. An experiment, first performed in 1979 by Hans Dehmelt and others, caused a single barium atom, trapped in a magnetic field, to reveal its presence to eyesight by

---

[7] These thoughts about the *Iliad* are worked out in relation to Aristotle's account of tragedy in my article "Tragic Pleasure" in *The St. John's Review*, Vol. XLIII, no. 1 (1995), pp. 21–37.

a starlike blue glow.[8] In both these examples, being is understood as existence. As Anselm says in Chapter II of his *Proslogium,* "it is one thing for an object to be in the understanding, and another to understand that the object exists."[9] The latter mode of being, he argues, must be conceived to be greater. Existence is ordinarily attested to by the senses, either directly or through such evidence as a photographic record or the testimony of witnesses. Anselm claims that, in the case of God alone, it is possible to move by thinking from an inadequate conception of what God would be to the certain knowledge that God exists. His argument will not be examined here. Aristotle does not argue from the thinkable to the existent, but uncovers another way of being that transcends and governs both the thinkable and the existent.

Aristotle's study of being is a working out of the cryptic claim in Plato's *Republic* (509B) that the good is beyond being, and is responsible for both the being of what is and the being-known of what is known. Socrates warns that one cannot see well the greatest of learnable things by means of his images of the sun, the divided line, and the cave, but only by "another, longer way around" (504B). Aristotle's *Metaphysics* is that longer way around,[10] not only in its general purpose of uncovering the greatest of knowable things, but also in its particular understanding of the dependence of being upon *forms.* The meaning of form is explained by Socrates in the passage we have just been quoting from. Besides the many beautiful or good things, he says, there is "also a beautiful itself and a good itself," the single "look" or form that is what each thing within the many *is* (507B). There is no "Platonic theory of forms."[11] The closest thing in the dialogues to such a theory is Socrates's brief flirtation, in the *Parmenides* (132B), with a claim about the place and source of the forms. He asks "might it not be

---

[8] Reported in *Physical Review A,* Vol. 22, no. 3 (1980), p.1137. (I saw a similar event on film on the Public Broadcasting System.)

[9] *St. Anselm: Basic Writings,* Open Court, 1962, p. 7.

[10] I do not mean that it is the only such way, or that Socrates had in mind or would agree with everything Aristotle says in the course of it, but only that it answers Socrates's call to address his questions without the shortcuts of his images.

[11] Books XIII and XIV of the *Metaphysics* are intended to refute certain hypotheses about the forms made by Plato and some of his students in the Academy. Aristotle's point is that no theory can be sustained by these speculations, which collapse as soon as any inferences are drawn from them.

that each of these forms is a thought, that has no business at all being present anywhere other than in souls?" By asking in return whether it is a thought of something or a thought of nothing, Parmenides firmly moves the form back from being a theory to being the topic of an inquiry, or else a tentative assumption for the sake of argument. (Cf. 135 B–C.)

There is no question that we get at the forms of things by thinking, but to take that as the end of the story dodges the issue of what they are. Thoughts come in many varieties, and beyond that, there is the question whether thinking is self-contained, related only to its own content, or whether it is one of the means by which distinct beings, knower and known, can come into relation. In the former case, the forms would be dependent on the act of thinking; in the latter, thinking would be dependent, at least in part, on beings. We are usually willing to grant sense perception the possibility of being related to things other than its own contents, but we shy away from being consistent with this open attitude when it comes to thinking. It is this possibility that Aristotle confronts directly.

One of the ways Aristotle describes the task of the kind of inquiry carried out in the *Metaphysics* is that it should say with clarity what the forms are.[12] In the first book of that inquiry, he says that Plato posited the forms as unchanging beings, apart from perceptible things, in which the latter "participate," but that he left behind to be sought just what that participation might be (987b 7–15). To speak of participating in a pattern is to make a poetic metaphor empty of content, since it leaves unaddressed the question of what is at work in this relation (991a 21–24). This is not a rejection of Plato's teaching, nor is it the adoption of a doctrine: the teaching determines a task and a goal, which Aristotle undertakes, but it does not give him a theory to work from. As Aristotle puts it in Book III, "we say that the forms are causes and have being in their own right . . . but in a number of ways these things are hard to swallow" (997b 3–6). This use of the first person plural is the signal for a historical scholar to start dismantling the text,[13] but *we*, if we are willing to embark upon a dialectical inquiry in a dialectical

---

[12] *Physics* 192a 34–36.

[13] Werner Jaeger, *Aristotle: Fundamentals of the History of his Development,* Clarendon Press, Oxford, 1934, p. 171. See Section II of this introduction.

spirit, can see it as linking not some conjectured early community of "Platonists," but Aristotle and ourselves, in an enterprise that is in no way historical. We need not locate ourselves within a collection of beliefs imagined to belong to Aristotle before or after he left Plato's Academy, or to the philosophizing of ancient or modern times. We are free to take seriously any preliminary account of the way things are, and see where it might take us.

It should be clear, though, that, for Aristotle as for any philosopher, the study of being can get at things only by way of thinking about them. If following his study uncovers anything that is new to us, that can only be because he has made us think anew. Anselm's argument has something of the feel of producing a rabbit out of a hat, since the certainty of the existence of an unacknowledged being is supposed to come, for even the most resistant reader, out of close attention to a thought. Something similar is at stake in Aristotle's inquiry, but it lacks any claim to coerce belief. It is true that Aristotle says that some people who are mistaken are in need not of persuasion but of "brute force," but these are people who deliberately argue sophistically, and the force is directed at demolishing their arguments (1009a 17–22). Even this departure from the dialectical approach occurs only to head off those who block the way into rational inquiry from the outset, by refusing to avoid self-contradiction, and then not to compel them to assent to any definite opinion, but only to accept that what they say must *mean* something (1006a 18–23). That is the ultimate foundation of dialectic: that beliefs can be articulated, and that those articulations mean something.

In Plato's *Gorgias* a series of accomplished and aspiring rhetoricians take turns congratulating themselves on their ability to make words do whatever they want them to. Socrates repeatedly shows them that their words commit them to more than they ever intended, while he claims that his own purpose is to follow, wherever it leads, the course laid out by speech itself. This has nothing whatever to do with investigating the meanings of words or of linguistic structures. The rhetorician is mistaken in thinking that human will is dominant over human speech, but that does not mean that we must be ruled by grammar and semantics. Language is an indispensable but flexible instrument of thinking. In order to think well about being, Aristotle remakes the Greek language in a number of important ways. These ways will be discussed in some detail in the third part of this introduction. For the moment, the reader should understand that any attempt to follow where Aristotle leads will oblige us to grant some flexibility to the ordinary uses of our own language.

At a crucial place in the *Metaphysics*, when all the preliminaries are done and the central argument is about to begin in earnest (1028b 3–5), Aristotle claims that the question underlying all philosophic endeavor, past, present, and future, is: What constitutes, and what is responsible for, the thinghood of things? He argues that the thinghood of any thing is what it keeps on being in order to be at all. This is what is meant by a thing's form, but now the notion of form must be allowed to develop fully. In Plato's *Republic,* as we noticed above, the form comes to sight first as the unity in speech that corresponds to a many that is perceived (507B), but the image of the divided line makes the form the original being of which the concrete particular is only an image (510B); the word abstract cannot apply in any way to the form, since the form is fuller than and independent of the concrete things. Aristotle follows Plato in concluding that the form is what is responsible for the thing's being as it is, and that it therefore cannot be a "universal" or mere idea that depends on thought or speech (1040b 28–1041a 6). Through most of Book VII (in Chapters 4–6 and 10–16), Aristotle argues *logikōs,* that is, from the evidence of thought and speech, but in the last chapter of that book, he turns to the direct evidence of perception that some things are genuine wholes and not mere heaps. As the one cause of many particulars, the form cannot itself exist as a concrete particular, but since it is the *cause* of the thinghood of those things, the form cannot be an abstraction or a universal. Similarly, the form is not one of the parts of a thing, nor the sum of its parts, but is responsible for its wholeness, and thus also for its parts. (1041b 7–31) It is this analysis of thinghood as form that shows that there must be a way of being that takes precedence over both what exists and what is merely thought, without being the same as either, and that governs both part and whole, without being part or whole.

Everything Aristotle accomplishes in the *Metaphysics* is achieved by hard thinking, but the result is not about thinking and certainly not about language. That result is the setting before us of a third way of being that neither perception nor customary habits of thought can prepare us for. Aristotle has not tried to make a mixture of the existent particular with the thinkable universal, but demands of us that we go beyond the limits of our deepest opinions about what can be. By the experience of sense perception, the highest kind of being is that of a thing that maintains itself as a unity with an enduring identity; such a thing is made out of material, but that material is made into some form. The material, which might have come to be many other things, cannot account for such a thing, but if its form is taken as a universal, a motionless object of thought, that form cannot account for the thing either. The only way to make sense of the world of experience is to open up the possibility of a new way of being that could belong to form, active and causal enough to be responsible for the way things are, yet stable and independent of those things. This way of being is what Aristotle calls *being-at-work*. It is evident in the things around us as life, and in the cosmos as a whole as the patterns of heavenly motions, but it belongs properly and directly only to forms. As it belongs to forms themselves, such being-at-work is beyond any ordinary and familiar notion of being; as the source and foundation of all other being, it is being as being, in its own right; and as the source of all wholeness and order, it is the good.

Books VII through IX establish the idea of being-at-work as the manner of being of the forms, which are in turn the causes responsible for the being-as-they-are, and the being at all, of animals and plants. This portion of the *Metaphysics* is linked to the culminating Book XII by the twofold bridge of Book X, an independent treatment of the topic of wholeness, and Book XI, a compilation of arguments from the *Physics* and earlier books of the *Metaphysics* designed to show that all apparent disorder in the universe is incidental to an overarching order. The common source of the unity of each living thing, and of the ordered cosmos of moving bodies, is disclosed in Book XII. The primary dialectical contribution of Book X is its opening argument that what is *one* most of all depends on an indivisible act of thinking. This converges with the thread of Book XI in Book XII, Chapter 7, where Aristotle argues that the motionless cause of all motion must be an unchanging object of thinking. Thus the whole dialectic of the *Metaphysics* calls for a being that is at once thinker and thought, upon which all that *is* depends, and which Aristotle calls god. As activity it is the active and productive intellect, while as content it is the totality of forms, separate and unchanging, defining the array of beings possible in this world. In this account, all being is dependent upon an act of thinking, but does not thereby become mere "ideas" or universals; that act of thinking does not make up a world, but secures the identity and stability, and thus the very being, of the existing things in it. At the end of Book III, Chapter 5 of *On the Soul,* Aristotle says of this same being, "without this, nothing thinks." The analogous conclusion of Book XII of the *Metaphysics* could be formulated as "without this being-at-work, nothing *is.*"

All thinking about the way things are is a quest for what is permanent, or at least enduring, among, within, or behind them. For Parmenides and the other Eleatic thinkers, this meant a rejection of all motion as illusory, and a trust in intellect alone as the way to truth. For Thales and the other Ionian thinkers, it meant the positing of some underlying material as the only stable being amid the variety of appearances. Thus the beginning of our philosophic tradition was a search for the inert, and this impulse has never gone away. For Aristotle, such an account of things is never explanatory, since the inert cannot be responsible for its own changes and transformations. There must always be a second kind of source at work in things, and this realization reopens the philosophic task (984a 17–23, 984b 9–18). Even in the most recent phases of the quest for wisdom, there is a persistent return to a faith in the inert among many thinkers, though it is followed always by the sort of failure that Aristotle claims must be forced by the truth itself. The physicist Newton identifies body with inertia, but then finds that body itself must also be a power that sets other bodies, and itself, in motion. The chemist Dalton identifies the elements with atomic bodies so hard that they can suffer no change, but the discoveries of Rutherford compel an attempt to picture the atom as a stable system of bodies in motion. The biologists Watson and Crick identify the species of living things with molecules whose structure can be changed only by the slow effects of random mutation, but Barbara McClintock discovers that those molecules actively rearrange themselves, and the work of John Cairns in the last few years demonstrates that they even direct their own mutation.[14] These discoveries all remain mysterious for our contemporary sciences, since they confine themselves to a way of

---

[14] The primary references for these claims are: Isaac Newton, *Philosophiae Naturalis Principia Mathematica,* 1687; John Dalton, *A New System of Chemical Philosophy,* 1810; Ernest Rutherford, in *Philosophical Magazine,* Vol. 211 (1911), p. 669; James Watson and Francis Crick, in *Nature,* Vol. 171 (1953), p. 737 and p. 964; Barbara McClintock, in *Cold Spring Harbor Symposium on Quantitative Biology,* Vol. 16 (1952), p.13; and John Cairns, in *Nature,* Vol. 335 (1988), p. 142.

explanation that can never succeed. The very attempt to posit inert-
ness in the cosmos, the atom, and the gene has disclosed that each of
those things can only be at all if it is a being-at-work. What Aristotle
describes as happening to the earliest seekers after truth (984a 18–19,
984b 9–11) has happened again among the most recent, as the truth
itself that they have both sought and denied has unfailingly forced
itself into their path: being is being-at-work.

This claim, in all of Aristotle's works other than the *Metaphysics*, is
an ultimate postulate and a guiding vision. In the *Nicomachean Ethics*
it lies behind the conception of moral virtue as the fully human life,
but there are other lives. In *On the Soul* it permits an understanding
of life itself, but there are nonliving things. In the *Physics* it is the
comprehensive background of motion and change in all nature, but
nature may not be the whole of things. It is the *Metaphysics* in which
the account of the whole must come to completion, and in which being-
at-work moves from being postulated to becoming the ultimate topic of
inquiry. Because that inquiry is dialectical, nothing so crude as proof is
attempted, but rather there is a sharpening and converging of a number
of ways of getting at and behind the being of all sorts of beings. That
is why, of all Aristotle's writings, the *Metaphysics* is made up of the
largest number of distinctly composed parts, but is held together by
the tightest internal unity. It must, if it is to succeed, draw together and
illuminate every kind of study, and lead the inquirer to the threshold
of wisdom. Across that threshold, the activity would be contemplative
rather than discursive.

II. Ways of Interpreting
The Distorting Lenses of
Modern Philosophy and Commentary

There is much that is valuable in the current secondary literature
about Aristotle in general and the *Metaphysics* in particular. A brief
list of suggested reading at the end of this book points the way to
some of these writings, which can themselves continue to guide a

reader who finds any of them useful or enjoyable. But in nearly all the commentary on the *Metaphysics* written in recent centuries, what is good needs to be disentangled from a variety of mistaken assumptions about what needs to be interpreted and how that should be done. Some writers have not even hesitated to declare the work uninterpretable. David Lachterman has preserved, in Latin, the judgment of a nameless eighteenth-century scholar that Aristotle's *Metaphysics* is a "sprawling, formless, horrendous monstrosity, in which the light of day has been left behind (*adeptum est*)."[15] Paul Shorey displays the same dismissive certitude, if not the same passionate frustration, in 1924, declaring that "the *Metaphysics*, as it stands, is a hopeless muddle in which no ingenuity of conjecture can find a certain order of thought," explaining in 1927 that "it will never be possible to clear up the confusion in Aristotle's *Metaphysics*, because Aristotle himself never cleared up that which existed in his own mind."[16] These are readers who write not to help others learn from Aristotle, but to explain their own lack of success in learning anything from him.

Much of the commentary written in this century has construed the task of understanding the *Metaphysics* not as a philosophic activity at all, but as a historical study. Beginning with the bias that the parts of the *Metaphysics* can not cohere in one simultaneous act of understanding, however complex its ordering, such interpreters conjecture that those parts can fit together only as stages in some sort of development. This has been a dominant approach to the *Metaphysics* since 1912, when Werner Jaeger published a study that applied to Aristotle's book the methods of a detective, drawing conclusions not from internal evidence but from facts outside the work itself.[17] Sir David Ross, in 1957, called Jaeger "the most brilliant Aristotelian of our time," and confessed to some pride that his developmental hypothesis had

---

[15] "Did Aristotle 'Develop'?," *Revue de Philosophie Ancienne* VIII, 1 (1990), p. 5. The quotation is three removes from its source. I am advised by Nancy Buchenauer, an experienced hand in such matters, that an error in the transmission of the verb, somewhere along the line, is not unlikely. If *adeptum* should be emended to *abreptum,* the light is not only left behind but snatched away, and if it should be *ademptum,* the light has died.

[16] *Classical Philology* XIX, p. 382, and XXII, p. 422

[17] *Studien sur Entstehungsgeschichte der Metaphysik des Aristoteles* (Weidmann, Berlin, 1912), incorporated in a later book translated as *Aristotle: Fundamentals of the History of his Development* (Clarendon Press, Oxford, 1934).

been anticipated in 1910 by an Englishman, Thomas Case, in the entry
on Aristotle in the eleventh edition of the Encyclopedia Britannica.[18]
Joseph Owens, with something more restrained but quite different
from pride, traces this sort of interpretation back to a different sort
of historical criticism found in an 1888 article by Paul Natorp, who
wanted to remove from the *Metaphysics* as not "genuinely Aristotelian"
those formulations which Case and Jaeger considered early and im-
mature.[19] Lachterman, in the article cited above, traces the historical
approach to the interpretation of the *Metaphysics* back another generation, to the French Hegelian Karl Ludwig Michelet, who won a prize
in 1836 by discovering unity and harmony amidst the appearances
of disarray and incongruity in the work.[20] This is yet a third way of
using the idea of historical development. Where Jaeger conjectured a
history of Aristotle's thinking, and saw him changing his mind, Natorp
imagined a subsequent history involving other people, and saw later
hands corrupting his text; Michelet must have seen contradictory formulations being generated in a historical process within the thoughts
themselves, and overcome in the grand sweep of the emergence of the
Idea.

There are serious objections to all three of these approaches. The
doctrines of Hegel are not universally accepted, and ought not to be
projected onto an author who had no acquaintance with them. The
appeal to corruptions of the text, wherever one is unable to make sense
of apparent contradictions, is little more than a counsel of despair. And
the conjecture that Aristotle changed his mind is the most popular
proposal of the three for the same reason that it is the least worthy of
credit: one can make it tell any story one pleases. Jaeger tells us that
Aristotle started out believing in a Platonic theology and later rejected
it in favor of a scientific empiricism. Now the first of these objects of
belief means almost nothing at all, and the second is nothing that could
conceivably mean to Aristotle anything like what it conveys in Jaeger's
notion of it, which derives from a nineteenth-century conception of

---

[18] *Proceedings of the British Academy* 43; reprinted in *Articles on Aristotle,* edited by
Barnes et al., Duckworth, London, 1975, Vol. I, pp. 1, 4.

[19] Joseph Owens, *The Doctrine of Being in the Aristotelian Metaphysics,* Pontifical
Institute of Mediaeval Studies, Toronto, 1951, pp. 35, 53–55.

[20] "Did Aristotle 'Develop'?" (cited in footnote 15), pp. 8–9

scientific method. Jaeger's own spirited but superficial grasp of a perennial conflict between science and religion is the pattern which Aristotle is cut to fit. But there is more than one way to skin this cat, and the German scholars Gohlke and Wundt found a way to tell the same story in reverse: emerging from the Academy as an empirically inclined materialist, Aristotle gradually came around to an acceptance of his teacher's theology. And if history's arrow can be reversed it can also be bent into a circle, so the Italian scholar Oggioni tells of an Aristotle who began his thinking within Platonic theology, departed from it, and finally returned to it in a revised form, all three stages being mixed together in the *Metaphysics*.[21] Someone once told me that there was a book that claimed the *Metaphysics* was not written by Aristotle at all, but by two of his students, who disagreed with each other; I think he was making a joke.

One work of twentieth-century scholarship stood against the tide of this frivolous descent into questions of historical "development" and reconnected contemporary interpretation of the *Metaphysics* with the many centuries of serious commentary upon it. This work is Joseph Owens's splendid study *The Doctrine of Being in the Aristotelian Metaphysics*. But Owens understands that we cannot grasp Aristotle's meaning with all the baggage of our usual presuppositions intact. To begin to confront his thinking, one must overcome or hold back from adopting the deepest tendencies underlying all the thinking of modernity. "Aristotle's procedure," he writes, "is to let the *things* speak for themselves. He waits for them to reveal their own inner nature . . . It is not a process of 'making our ideas clear,' or of bringing order by means of ideas."[22] By the canons of virtually all philosophic thinking of the last four hundred years, what Aristotle does ought to be impossible. At the origin of the modern turn to subjectivity,[23] Descartes asked, in the second of his *Meditations* as in the fourth part of his *Discourse on*

---

[21] These permutations of Jaeger's tale are reported by Giovanni Reale in *The Concept of First Philosophy and the Unity of the Metaphysics of Aristotle* (SUNY Press, 1980, pp. 5–7).

[22] See *Doctrine of Being* (cited in footnote 19), p. 131.

[23] This might be described more accurately, if obscurely, as a shift in the meaning of subjectivity. In medieval texts, the "objective" aspect of a thing is its relation to a perceiver or knower, while its "subjective" aspect is what the thing is in itself, independent of any such relation. When Descartes uses these words at all, it is in their older sense, but he paves the way for shifting the meaning of "subject" to the

Method, "What am I?" and answered "I am a substance the whole essence or nature of which is only to think."[24] This "I" intends not Descartes alone but anything that can say "I." As judged from this standpoint, not even Aristotle can know anything but his own mental states. The prevailing opinions of a modern age are then projected onto a thinker who had explicitly denied them, as Owens shows (pp. 128–129).

When, in the twentieth century, certain thinkers follow up the earlier subjective turn with a new "linguistic turn," interpreters of Aristotle turn him again, to suit the current taste. Discussing the first book of Aristotle's *Physics* a few years after Owens wrote, Wolfgang Wieland says "What Aristotle undertakes in this context, in fact, is simply an investigation into what presuppositions we have already made if we speak of natural things and events . . . includ[ing among other such presuppositions] those structures which we accept simply through talking in a particular language."[25] In his effort to take Aristotle seriously in Aristotle's own terms, Owens takes as his chief opponent (p. 132) R. G. Collingwood, for whom a theory of knowledge is either idealism or else "naive," "low-grade," and "based upon . . . stupidity." In the final footnote of his book, some 435 pages after Owens first refers to Collingwood, he tacitly pronounces him refuted by echoing those epithets. But Collingwood saw himself as doing battle with a new, self-described "realism," the British analytic philosophy that admired and sought to imitate modern physical science, that severed philosophy from life, and that later encouraged the turn from being to language.[26] Collingwood, who recognizes the dignity and power of thinking, is certainly closer to the spirit of Aristotle than is this recent trend.

Genuine philosophic activity demands more effort than we customarily make, both in sustained concentrated thinking and in the

---

self. The sense of the words that is now so familiar to us was permanently fixed by Kant.

[24] See his *Philosophical Writings* (cited in footnote 4), Vol. I, p. 127 and Vol. II, p. 19. The translation is the earlier Haldane-Ross version, as corrected by Richard Kennington.

[25] Kant Studien 52 (1960–61), reprinted in *Articles on Aristotle* (cited in footnote 18), Vol. I, p. 132.

[26] R. G. Collingwood, *An Autobiography,* Oxford University Press, 1970, pp. 47–52. Collingwood's death in 1943 spared him from seeing the full extent to which philosophy could be made into a "silly and trifling game."

sustained suspension of all our comfortable assumptions about things. There is always considerable appeal, then, in a theory that says a philosopher is just changing his mind irrationally, or just misunderstanding or misusing his language. Jaeger made it respectable to substitute casual historical conjecture for philosophic interpretation; analytic philosophy, which has gone by many names, has often sought to replace philosophic thought with the study of language or logic. Collingwood dates the rise of the latter from 1912 (the very year Jaeger's thesis appeared in print), and from such people as Bertrand Russell, but its roots go back to Thomas Hobbes and David Hume. Writing in 1651, Hobbes finds that everything in Aristotle's *Metaphysics, Physics,* and *Ethics* depends on "Terms . . . [that] are no names of things," "empty names," and "mere words, and in this occasion insignificant."[27] A century later, Hume decides that "when you pronounce any action or character to be vicious, you mean nothing, but that, from the constitution of your nature you have a feeling or sentiment of blame from the contemplation of it."[28] The thinkers admired in the past were deluded souls, ignorant that their highest achievements were merely insignificant speech or emotive language.

In Plato's *Phaedo* (89C–90D), Socrates diagnoses any antiphilosophic turn of this sort as "misology"; it arises especially, he says, in debaters, who, because they can somehow or other refute anything, conclude that there is "nothing wholesome or stable in speech." But, Socrates concludes, "it would be a pity if . . . one were not to blame himself and his own lack of skill . . . but deprived himself of the truth and knowledge of beings." One ought at least to try to meet Aristotle on his own ground. It would be important to learn whether his arguments have a power that is not limited by history and language. The claim is made often these days that Jaeger's approach to Aristotle is out of fashion, and I am told that the linguistic approach to philosophy has become less rigid and more open-minded, where it is still to be found at all. If these assertions are true, then even among his professional interpreters, Aristotle can be accorded a new and open hearing. This translation and account of Aristotle's *Metaphysics* is addressed

---

27 Thomas Hobbes, *Leviathan*. These and similar criticisms are found throughout Chapter 46.

28 David Hume, *A Treatise of Human Nature*, Book III, Part I, Sec. 1. (Oxford, Clarendon Press, 1978, p. 469.)

to all those who are willing to refrain from judging the enterprise in advance, and want to follow where Aristotle leads. This requires a willingness, above all, to recognize and acknowledge that we do not know the limits of what can be. It requires also, in particular, a refraining from the habit of resorting to some vague notion of "concepts" to explain everything that we encounter by thinking.

Aristotle makes use of language to point beyond what depends on language. Medieval authors distinguished between the first intention of an act of understanding, and its many possible second intentions. If I say "beauty" is a thought, or "beauty" is a word, I am referring to second intentions, while the first intention would be the thing itself that is thought or spoken about. Jacob Klein has analyzed the transition from ancient to modern mathematics, by which *signs,* understood as pointing beyond themselves, were transformed into *symbols,* having meaning only within a self-contained realm of symbols such as algebra. In Descartes's *Rules for the Direction of the Mind,* the understanding of number as a multitude of units is replaced by the intellect's "own conceiving of that 'multitude of units,' that is, the 'concept' of number as such." When this "'second intention' is grasped *with the aid of the imagination* in such a way that the intellect can, in turn, take it up as an object in the mode of a 'first intention,' we are dealing with a *symbol.*"[29] This shift in mathematics was a world-changing event, extending to everything thinkable. If, when Aristotle identifies being with form in Book VII of the *Metaphysics,* we have any temptation to replace forms with "concepts," we are too firmly embedded in the tradition stemming from Descartes to hear Aristotle's questions.

Thomas Aquinas understood knowledge as "intentional being," the presence in the mind, by way of their effects, of the things which have their proper being outside the mind, for "that which is in something does not follow that in which it is . . . Similarly, truth, which is caused in the mind by things, does not follow the estimation of the mind, but the existence of the things."[30] But on the other side of the looking glass, Edmund Husserl makes the "intentionality of consciousness"

---

[29] Jacob Klein, *Greek Mathematical Thought and the Origin of Algebra,* M.I.T. Press, 1968, p. 208.

[30] Thomas Aquinas, *The Disputed Questions on Truth,* Q. I, Art. I, reply to objection 3, in *Selections from Medieval Philosophers,* Scribner's, 1930, Vol. II, p. 174 and note on p. 466.

a purely internal feature: "The 'object' of consciousness ... does not come into the process from outside; on the contrary, it is included as a sense in the subjective process itself—and thus as an *'intentional effect.'*"[31] Putting together the Thomistic and Husserlian meanings, we can see that some sort of intentionality is inevitable in thinking, whether as a sense within consciousness that something inner has an outer reference, or as a capacity in the knower to come into relation to things that are truly external. The notion of forms is compatible with either of these attitudes. While it is certainly wrong to think that Aristotle assumed the forms to be mere concepts or linguistic artifacts, Owens goes too far in the opposite direction in claiming that Aristotle simply takes over into the *Metaphysics* an "explanation of cognition" (p. 128), worked out in *On the Soul,* that has established all known things as independent beings. That the forms have some sort of independence of the human knower is undoubtedly a conclusion of *On The Soul,* but that does not establish them as altogether independent beings. They might still be only thought-dependent entities, secured by the thinking of a being that is not human or perishable. The relation between the forms and the active intellect itself is explored not in *On the Soul* but in Book XII of the *Metaphysics.* The forms are uncovered in various ways in a number of Aristotle's writings, but everything said about them outside the *Metaphysics* is in some way provisional. What is said about them in the *Metaphysics* has been sketched in the first part of this introduction, but we need to examine it in more detail. Those details, however, will be accessible to us only if they have not been lost in the translation.

---

[31] Edmund Husserl, *Cartesian Meditations,* Nijhoff, The Hague, 1969, p. 42. On the history of the word intentionality, see Owens's long footnote in *Doctrine of Being* (cited in footnote 19), pp. 133–135.

III. Ways of Translating

How and Why this Version Differs from Others[32]

We cannot give an accurate summary of Aristotle's conclusions about the way things are unless we have some way to translate his characteristic vocabulary, but if our decision is to follow the prevalent habits of the most authoritative interpreters, those conclusions crumble away into nothing. By way of the usual translations, the central argument of the *Metaphysics* would be: being qua being is being per se in accordance with the categories, which in turn is primarily substance, but primary substance is form, while form is essence and essence is actuality. You might react to such verbiage in various ways. You might think, I am too ignorant and untrained to understand these things, and need an expert to explain them to me. Or you might think, Aristotle wrote gibberish. But if you have some acquaintance with the classical languages, you might begin to be suspicious that something has gone awry: Aristotle wrote Greek, didn't he? And while this argument doesn't sound much like English, it doesn't sound like Greek either, does it? In fact this argument appears to be written mostly in an odd sort of Latin, dressed up to look like English. Why do we need Latin to translate Greek into English at all?

At all the most crucial places, the usual translations of Aristotle abandon English and move toward Latin. They do this because earlier translations did the same. Those earlier translators did so because their principal access to Aristotle's meaning was through Latin commentaries. The result is jargon, but that seems not to make most of the professional scholars uncomfortable; after all, by perpetuating such inaccessible English texts, they create a demand for interpreters that only they can fill. I have criticized the current state of Aristotle-translating at length in the introduction to a recent translation of my own.[33] There it seemed necessary to explain to those familiar with other translations

---

[32] The choices I have made for translating the principal words and phrases of Aristotle's vocabulary are gathered together and explained in the Glossary at the end of this introductory material.

[33] *Aristotle's Physics: A Guided Study,* Rutgers University Press, 1995 and 1998.

the many departures they were about to encounter. Here a briefer justification may suffice: My aim is to give you in translation an experience
as close as I can make it to reading the original. The original is written
not for specialists but for generally educated people of any sort who
are willing to think hard. Where Aristotle exploits the resources of the
Greek language to capture his meaning, the translation will be in bad
English; where he departs from Greek usage to coin new words and
novel ways of saying things, the translation will be in worse English.
From the point of view of a classicist, a good English translation of
a classical author is one that finds, for every word or phrase in the
original, some equivalent expression that reads smoothly in our language. This may be a good practice with some kinds of writing, but
philosophic meaning cannot be captured in habitual uses of language.
The point of view of a professional philosopher may, however, pay
too much heed to the linguistic choices that have become habitual in
modern philosophy and in the secondary literature, at the expense of
faithfulness to the original.[34]

The translation presented here is governed instead by the needs
and purposes of the careful student. The feel of the classroom is everywhere in Aristotle's words, and has not been disdained here. In no
case will the translation have a learnèd style; like the original, it will
rely on the most ordinary contents of its own language, achieving the
distinction its topics demand by recombining those simplest elements
in unaccustomed ways. Some explanation will be necessary in footnotes, for the sake of cross references, to fill in what is absent where
the text is elliptical, and to highlight important steps in the inquiry,
but the translation is intended to reduce the amount of commentary
needed. Wherever I think I know something that will enrich the reading of the text, I have noted it, but you should never take these notes as
authoritative if they conflict with any interpretation that seems better
to you.

Three outstanding readers of Aristotle led me to see that a new
way of translating him was necessary and possible. Jacob Klein, in

---

[34] This concern may be seen in the scruples that burdened Marjorie Grene, mentioned
in the last sentences of the book included in the brief list of suggested reading given
at the end of this volume. Her renderings of two of Aristotle's most important terms
do not follow her own preferences.

his extraordinary brief essay "An Introduction to Aristotle,"[35] helped me begin to encounter Aristotle's thinking directly and genuinely. The same service had been performed for Klein a generation earlier by Martin Heidegger's lectures on Aristotle.[36] In various writings, not primarily about Aristotle (as, for example, in the first few sections of *What is a Thing?*), Heidegger claims that the fundamental question of philosophy concerns the thinghood of things. This thought, when compared with "the substance of substances," began to unlock for me a way of getting at one of Aristotle's central ideas in intelligible, though unusual, English. Finally, Joseph Owens has exposed the absurdity of translating Aristotle's word *ousia* as "substance" and the inadequacy of translating his phrase *to ti ēn einai* as "essence."[37] By Owens's account of the history of this use of the word "substance," it stems from a Latin word conceived as *negating* what Aristotle meant by *ousia* in the *Metaphysics.* It is much as if one were to take Thomas Aquinas's use of the word "objective" in a Kantian sense, or his use of "intentionality" in a Husserlian sense. Owens knows the Latin tradition well enough to escape from its distortions and mistakes.

But when Owens attempts to replace "substance" and "essence" with more accurate substitutes, he is not successful. To replace the latter he translates *ti ēn einai* what-IS-being, explained as meant to convey timelessness. Here it is rendered most fully as "what something keeps on being, in order to be at all," or more often as "what it is for it to be." This follows Owens's general interpretation that the main structure of Aristotle's phrase is not the articular infinitive *to einai,* complemented by *ti ēn* (the being-what-it-is of a thing), but a novel coinage of Aristotle's, an unambiguous transformation into the progressive aspect of the *ti esti* that Socrates was always in quest of

---

[35] In the anthology *Ancients and Moderns* (Basic Books, New York, 1964) and in Klein's *Lectures and Essays* (St. John's College Press, Annapolis, Md., 1985). The latter is still in print, but less likely to be found in libraries.

[36] The lectures Klein heard are now available as the first part of Heidegger's book, *Plato's Sophist* (Indiana University Press, 1997). Two later lecture courses can be found, one in the article "On the Essence and Concept of *Phusis* in Aristotle's *Physics* B,1" in the book *Pathmarks* (Cambridge University Press, 1998), p. 183, and another in the book *Aristotle's Metaphysics Θ 1–3* (Indiana University Press, 1995).

[37] See *Doctrine of Being* (cited in footnote 19), pp. 138–147 and 180–188.

(the what-it-enduringly-is).[38] It differs from Owens's version in taking the *einai* as an infinitive of purpose ("in order to be at all"); with the infinitive so understood, Aristotle's phrase captures the meaning of the necessary-and-sufficient being of a thing, excluding not only what a thing is in a transient and incidental way, but also what it is in a partial and universal way, that is insufficient to make it what it is.[39]

Thus, the way I understand *to ti ēn einai* departs from, but is rooted in, Owens's understanding of it. The same is true of my rendering *ousia* as "thinghood," when it is used in a general sense, and as "an independent thing" when it is used of singulars.[40] I have heard two sorts of criticism of my use of the word thinghood in Aristotle's *Physics*. The one sort, that it occasions laughter or embarrassment, is a general instance of Heidegger's observation in *What is a Thing?* that philosophy is that at which thoughtless people laugh. Let the laughter or embarrassment subside, and then judge the meaning carried by the word, both on its own and in its context, on its merits. The other sort of criticism regrets the fact that thinghood is not as closely related to being as *ousia* is to *to on*. I too regret it, but when one cannot have everything, one must make choices. It seems probable that *ousia* is simply the participle *on* plus the noun ending *-sia*, with the nu-sigma rule applied to make it not grate on Greek ears.[41]

---

[38] Verbs in ancient Greek are distinguished by aspect as progressive, aorist, and perfective. Only in the past tense are there distinct forms for all three. Hence the mere shift from *esti* to *ēn* highlights the progressive sense of the verb, and what is ongoing about being some type of thing.

[39] For instance, a human being is always an extended body, but to be an extended body is too general a characteristic to make something a human being. Aristotle's phrase, by the mere juxtaposition of a progressive finite form with the infinitive of *to be,* elegantly and compactly finds the exact level, between the too-particular and the too-general, that expresses what something is.

[40] In Greek, *ho anthropos* can refer either to humanity or to some single human, but we cannot say in English that an elevator can hold six humanities. The duality by which thinghood in its general sense refers both to the perceived bodiliness of singular things, and to the source of that very bodiliness in something imperceptible and nonbodily, must be justified by Aristotle's own argument in Book VII, Chs. 2–3.

[41] All published discussions of the word *ousia* known to me derive it from the feminine participle of to be, *ousa,* plus the more common noun ending *-ia*. Abraham Schoener has pointed out to me this likelier derivation. Smyth's *Grammar,* in Sec. 840 a, says that *-sia* is used only with verbs ending in *-azō*, but then derives *euergesia* from *euergetēs.* Either version would make it formally equivalent to beingness or beinghood, but the meanings of words depend on more than such formal relations.

The word *ousia* is used by Aristotle in two ways, and "beingness" conveys neither of them. Lassie is an *ousia*, and the *ousia* of Lassie is dog. To call her a beingness makes no sort of sense in English; she is a being, but this is the natural translation of *on*, which Aristotle uses more widely, for anything at all that is, including the color white. But not even in its general employment is the meaning of *ousia* captured by beingness. Lassie's beingness could mean the fact that she is, or it could ask in general about her way of being, but *ousia* in this usage names only one particular way of being. It is the first of the eight ways of attributing being (the so-called "categories") listed in the *Metaphysics*,[42] but while it is called *ousia* in the second of those lists, in the first it is called the *ti esti*, the "what it is" of something. Beingness does not match up with the what, just as being-what-it-is does not have the same meaning as what-it-is-for-it-to-be. Lassie's being a dog is not the same thing as dog, and the latter is what she is.

One way that ancient Greek differs from English is in having no word for thing, in the emphatic sense of the English word. There is a word (*chrēma*) that means a thing used up or consumed, another (*pragma*) that means a thing of interest or concern or a thing done, and a third (*hekaston*) that refers to each instance of a general kind, but most often things in any sense are signified by the neuter ending on an adjective or article. None of these ways of speaking singles out the emphatic character of that which stands on its own and can be pointed to, which is thus independent of what surrounds it as well as of what apprehends it, but the combination of those two notions (*tode ti te kai chōriston*) is exactly the way Aristotle characterizes *ousia* (1017b 27, 1029a 27–28). In so doing, I submit, he names the independent thing in its thinghood. Owens would have us think of this as "the entity in its entity" (pp. 149–154), but it is instructive to see that, in the argument that precedes this suggestion, Owens himself, spontaneously reaching for a way in English to denote that which is independent of the knower, repeatedly chooses and italicizes the word *thing*.[43] This is a vigorous and effective use of the English language. It is so simple and straightforward that no one needs to have it explained. This is exactly

---

[42] The list at 1017a 23–29 is repeated at 1068a 8–10 in different words and with time, or the "when," omitted.

[43] See *Doctrine of Being* (cited in footnote 19), pp. 112, 113, 127, 129, 130, 131, and 138.

the way Aristotle uses the Greek language. In speaking of thinghood I am imitating Aristotle by doing as Owens does, not as he says to do.

The question of the thinghood of things turns out, for Aristotle, to be the question of the manner of being of forms. The material of things is what might have come to be any of a variety of things, while the composite of form and material is an effect derived from some cause. The cause must in some way reside in the form. But while the composite thing exists as something perceptible and particular, and the material is separate from it only in thought and as a universal, some third way of being must belong to whatever is at work as causing the many things of any kind to be what they are.[44] In the first part of this introduction we have seen that this way of being is what Aristotle calls *being-at-work.* It is clear that Aristotle's word for being-at-work (*energeia*) must be at the center of his philosophic vocabulary, and that its meaning must be at the heart of all his thinking. What resources do we have to get at that meaning? Not only does Aristotle not define the word, he says in Book IX, Chapter 6 that it has no definition. Its meaning is at the limit of definition and explanation, and has nothing of greater defining or explanatory power to which it can be referred. Motion in the *Physics,* soul in *On the Soul,* and both form and being in the *Metaphysics,* are ultimately understood as kinds of being-at-work, and even the notions of virtue and character, in the *Nicomachean Ethics,* depend on it. Without a grasp of what he means by being-at-work, then, one cannot get hold of any important idea in any of Aristotle's writings. Aristotle invents a second word, being-at-work-staying-itself (*entelecheia*), converging with it in meaning, to sharpen and clarify his use of being-at-work, and he gives an array of examples in which we are meant to "see at a glance, by means of analogy," what it means (1048a 39).

---

[44] Joseph Owens (*Doctrine of Being,* pp. 390–395) seems to be alone in contemporary secondary literature in understanding and endorsing this. Edward Halper's excellent discussion of the issue in his book *One and Many in Aristotle's Metaphysics* (Ohio State University Press, 1989, pp. 229–244) confirms that this is still the case. Rogers Albritton, at the beginning of an article in the *Journal of Philosophy* (Vol. LIV, 1957, pp. 699–708), may speak for many in his complaint, "I doubt that Aristotle would have understood any better than I do the suggestion that a thing may be neither universal nor particular." Those few, such as Halper, who have seen how radical Aristotle's inquiry is, prefer to think of the form as *both* universal and particular, but Aristotle explicitly calls this a mistake (1086a 33–36), and distinguishes three sorts of contemplative knowledge that correspond to three ways of being (1026a 10–19 and 1064a 29–1064b 3).

The seeing that Aristotle calls on us to do demands that we come alive to the world in a way that can launch us toward philosophic questioning. No one can do it for us. But a thoughtless translation of the word that names what we are asked to see can deprive us of any chance to begin. In the usual translations *energeia* crumbles away to nothing as "actuality." Any hope of recapturing it through its near-synonym is lost, since these translations render *entelecheia* also as "actuality." Does this word give you any hint that Aristotle is responding to the call made to the philosopher by the Eleatic Stranger in Plato's *Sophist* (249 C–D) to approach being by thinking rest and motion together? Does it convey a stable condition that can be achieved only by ceaseless activity? Does it describe a motion that leads nowhere but back into itself? The Latin word *actualitas* may have performed those services for a reader of Latin, but its English cognate means nothing that remotely suggests them.[45]

The idea of being-at-work pervades Aristotle's thinking. His characteristic vocabulary emphasizes it everywhere. He chooses a common noun (*energeia*) built on the root *erg* that signifies work. He finds the same meaning in the common verb *echein* that means to be by continuing or holding on in some way, and attaches it to an adjective (*enteles*) that signifies completeness, to form the coinage *entelecheia,* which redoubles its meaning by punning on a common word (*endelecheia*) that means continuity or persistence. He remakes Socrates's favorite question *ti esti* (what is it?) by changing the verb to the past tense (*ēn*), in which alone its progressive aspect can be made unambiguous. And he chooses for the primary explanatory word in his ethical writings the noun *hexis* made from *echein,* to signify an enduring state of character that is also an active condition. His language pulses with this dynamic conception of being,[46] and guides a way of seeing the world

---

[45] Even Joseph Owens is content with the Latin root, translating both words as "act" (*Doctrine of Being,* p. 405), which is less bad but still carries none of Aristotle's positive meaning. Martin Heidegger sees what makes the crucial difference in that meaning, which is clearer in German than in Latin. (See p. 143 of *Aristotle's Metaphysics* Θ *1–3,* cited in footnote 36.) Jacob Klein is the only writer in English I know of who has had the wit to see how much rests on getting this right, and who has made it clear. (See the lecture cited in footnote 35: pp. 57, 64–65 in *Ancients and Moderns,* or pp. 180, 189–190 in *Lectures and Essays.*)

[46] One effect of translations that lose this meaning is the extensive literature in the philosophic journals in which analytic philosophers struggle to decide what sort of

as organizing itself into every instance of identity it presents. For the reader open to philosophy, these words of Aristotle are what Homer called "winged words," shafts that traverse the distance between people and lodge in the place where understanding is possible.

IV. Notes on this Volume

Running alongside the translation of Aristotle's text in the margins are the page numbers and column letters of the standard, two-column Bekker edition of Aristotle's works. The numbers following them correlate with the lines of both the Oxford Classical Text of the *Metaphysics*, edited by Werner Jaeger, and the older Oxford Text edited by W. D. Ross and still available in the two-volume version including commentary. Not only are the lines the same in the two versions, but to consult either one is to encounter the other within the critical apparatus at the bottoms of its pages, since Jaeger's emendations had been published in journals and monographs before Ross made his text. Ross's editorial judgment is the more trustworthy, since it is not driven by an interpretation, but I have occasionally departed from it when a variant reading made more sense to me. Ross's notes have been valuable to me as a source of references of all sorts. Joseph Owens's commentary, discussed above in this introduction, has been of more substantial help. Other worthwhile reading is listed at the back of this book, not as necessary preliminaries but as additional pleasures. The translation and brief footnotes are aimed at making the topic of first philosophy a matter for direct conversation between you and Aristotle.

---

verbs predicate "actuality" rather than motion. The earlier entries in this discussion focus their attention on "Aristotle's tense test" that distinguishes seeing from walking, in their view, on the ground that the former can be spoken of in both the present and perfect "tenses" at once, while the latter cannot. Eventually, it emerges into the debate that ancient Greek had a grammar different from that of English, and that Aristotle's distinction has nothing to do with tenses, and is not used as a test of anything. The discussion is immersed in questions of grammar and semantics, but in the wrong language, and only because translations have blocked access to the sources of meaning. My glimpse into this region of the secondary literature came through Mary Hannah Jones's excellent unpublished 1988 essay "Aristotle's Energeia-Kinesis Distinction."

V. Acknowledgments

My debt to St. John's College is beyond measure. As a freshman there more than thirty years ago, in seminars led by Robert Bart, I first encountered Aristotle as a living thinker. In graduate schools elsewhere, my studies of Aristotle took their origins from Jacob Klein's luminous lecture, "Aristotle, an Introduction." When I thought I saw a way to reinvigorate the translating of Aristotle's writings, St. John's supported my work with a two-year appointment to the NEH Chair in Ancient Thought. When I was engaged in that work, many colleagues and students gave me help and encouragement, and especially, during the last days of his life, J. Winfree Smith. When I had completed a translation of the *Physics*, Harvey Flaumenhaft made it possible for it to be published. And when I wanted to follow up with the *Metaphysics*, Dana Densmore and Bill Donahue welcomed me into the superior work of the Green Lion Press. All these teachers, coworkers, and friends, to whom I owe so much, have been part of the community of learning fostered by St. John's.

Joe Sachs
Annapolis, Maryland
Summer, 1998

A Brief Outline of the Argument of the *Metaphysics*

This summary ignores the complexity within each book of the *Metaphysics* in order to highlight the broad outlines of the work in its wholeness. The fourteen books group themselves into five main sections.

Preliminary Inquiries

Book I All those who have sought wisdom have tried to uncover the ultimate causes of things. Most have produced materialist accounts, which necessarily show themselves to be inadequate, but all attempts to describe a source of motion or a cause for the sake of which things are as they are have been rudimentary and explain nothing either.

Book II Any sequence of causes that succeeds in explaining anything cannot be infinite, but must lead to the primary instance of whatever it explains.

Book III A successful account of the causes of things must resolve a number of impasses. Principally, if a cause is to be knowable it must be a universal, and therefore cannot act in the way particulars do or have being in the way perceptible things do.

Book IV If there is a single knowledge that encompasses all things it must incorporate a knowledge of the first principles of knowledge itself, and especially of the law of contradiction. But that law governs speech and thought because it first of all governs beings. To be at all is to be something, to have a determinate nature and belong to a kind.

Book V Causal structures are reflected in the arrays of meanings that belong to words. When ambiguities are laid out, the complexities of causal relations may be discovered rather than imposed from some theoretical pattern.

The Central Argument

Book VI Being is meant in more than one way, but insofar as it refers to truth in thinking, or to anything that something is incidentally, its meaning is derivative from and dependent upon a more primary sense.

But being in its own right still belongs to things in a number of distinct ways, with which the study of being as being must begin.

Book VII Among the ways being is meant in its own right, thinghood is primary, since the other seven ways all signify being some attribute of an independent thing. Since the thing underlies its attributes, thinghood seems to be material, but material underlies form in a different way. The thinghood of a thing is disclosed not by subtracting its attributes, but by looking to the form that determines its wholeness as a separate being, standing on its own. The form is what it is for the thing to be, what it keeps on being in order to be at all, and as such belongs only to living things and to the cosmos as a whole. The articulation in speech of what it is for something to be defines it only as a universal, incapable of being separate or acting as a cause. As the cause of thinghood, form must be understood as having being prior to and independent of perceptible things. Being as being has been narrowed down to thinghood and then to the being of forms.

Book VIII Form is not an arrangement of parts but a being-at-work within material, and the unity in the form itself results from a being-at-work within it. Being as being is being-at-work.

Book IX The potencies in things are not mere possibilities, passive and externally related to something at work upon them, but inherent strivings that emerge into being-at-work in response to a cause in which that same being-at-work is already present. The perpetual renewal of being-at-work in living things and in the cosmos depends upon an everlasting and indestructible being, always at work. As the source of identity for whatever is, this everlasting being-at-work is the good of all things, and contact with it by the act of contemplation is the primary meaning of truth.

Bridge to the Conclusion of the Inquiry

Book X Wholeness in its most complete form belongs to that of which the thinking is one, and hence the cause of such wholeness rests ultimately upon an act of thinking.

Book XI There is much evidence of disorder among things. In the course of the *Metaphysics* and *Physics,* all such apparent disorder has been traced back to some sort of order on which it depends. Hence that order in turn must have some source.

Conclusion: the Source of Being

Book XII All things depend upon a being that is without material and always at work. As an act of thinking that never goes out of itself, it is the motionless cause of the circular motion of the cosmos; as the self-thought content of thinking it comprises the forms of all the beings that inhabit the cosmos.

Final Caution:
Misguided Approaches to the Source of Being

Book XIII Mathematical things are not sources of anything, either separate from or within perceptible things, and the forms are not mathematical things or universals.

Book. XIV The forms cannot exert causality as elements, by any analogy to generation, or as numerical ratios.

Glossary

The third part of the introduction discusses the ways in which this translation departs from others. This glossary sets forth the primary translation choices that constitute that departure.

Its first section lists, alphabetically, the Greek words that are discussed below, with the English translations that have, for the most part, been used for them here. In many cases, the translation chosen is followed in parentheses by the standard translation that is *not* used here. Some of these rejected translations, such as *perplexity* or *formula*, convey false impressions by wrong emphases; others, such as *habit* or *speculation*, are cognate to good Latin translations of the Greek, but have entirely different meanings in English; still others, such as *substance* or *induction*, are mistakes made long ago that have hardened into authority.

The second section of the glossary discusses all these words at some length, with reference to many of Aristotle's writings. The glossary thus supplements the introduction, and provides a way for you to orient yourself to Aristotle's primary vocabulary as a whole.

I. Greek Glossary

αἴθησις ... sense perception (*not* sensation)

αἰτία ... cause

ἀλλοίωσις ... alteration

ἀνάγειν ... lead back (*not* reduce)

ἀντίφασις ... contradictory

ἀπορία ... impasse (*not* perplexity, difficulty)

ἀρετή ... virtue

ἀριθμός ... number

ἀρχή ... source (*not* principle)

αὐτόματον ... chance

ἀφαίρεσις . . . abstraction

γένος . . . genus

δύναμις . . . potency (*not* potentiality)

εἶδος . . . form

ἐναντίον . . . contrary

ἐνέργεια . . . being-at-work (*not* actuality)

ἐντελέχεια . . . being-at-work-staying-itself (*not* actuality)

ἕξις . . . active state

ἐπαγωγή . . . example (*not* induction)

ἠρεμία . . . rest

θεωρία . . . contemplation (*not* speculation)

ἰδέα . . . form

καθόλου . . . universal

κατὰ συμβεβηκός . . . incidental (*not* accidental)

κίνησις . . . motion

κινοῦν . . . mover (*not* efficient cause)

λόγος . . . articulation (*not* formula)

μεταβολή . . . *See* motion

μορφή . . . form

νῦν . . . now (*not* moment)

ὁμωνυμία . . . ambiguity (*not* equivocation)

οὐσία . . . thinghood (*not* substance)

ποιόν . . . of-this-kind (*not* quality)

ποσόν . . . so much (*not* quantity)

πρώτη, πρῶτον . . . primary

πρώτη φιλοσοφία . . . first philosophy

στάσις . . . rest

στέρησις . . . deprivation (*not* privation)

τὰ μετὰ τὰ φυσικά . . . *See* first philosophy.

τέλος ... end

τέχνη ... art

τί ἐστι ... *See* what it is for something to be

τί ἦν εἶναι ... what it is for something to be (*not* essence)

τόδε τι ... this (*not* this somewhat)

τόπος ... place

ὕλη ... material (*not* matter)

ὑποκείμενον ... underlying thing

φύσις ... nature

χωριστόν ... separate

II. English Glossary

This is a slightly revised version of the glossary that appears with the translation of the *Physics*, based upon those passages in which Aristotle explains and clarifies his own usage. Bekker page numbers from 184 to 267 refer to the *Physics*; those from 980 to 1093 are in the *Metaphysics*.

abstraction (ἀφαίρεσις *aphairesis*) The act by which mathematical things, and they alone, are artificially produced by taking away in thought the perceptible attributes of perceptible things (1061a 29–1061b 4). Within mathematics, this is the ordinary word for subtraction. It is never used by Aristotle to apply to the way general ideas arise out of sensible particulars, as Thomas Aquinas and others claim. Its special philosophic sense is not Aristotle's invention; as often as not he speaks of "so-called abstractions." He uses the word in this special sense rarely, only in reference to the origin of mathematical ideas, and not always then; in the *Physics* he says instead that mathematicians separate what is not itself separate (193b 31–35).

active state (ἕξις *hexis*) Any condition that a thing has by its own effort of holding on in a certain way. Examples are knowledge and all virtues or excellences, including those of the body such as health. Of four general kinds of qualities described in *Categories* Book VIII, these are the most stable.

alteration (ἀλλοίωσις *alloiōsis*) Change of quality or sort, dependent upon but not reducible to change of place. One of the four main kinds of motion. Some things that we would consider qualities are "present in"the thinghood of a being, making it what it is, rather than attributes of it; change of any of them would be change of thinghood, rather than alteration of a persisting being (226a 27–29). The acquisition of virtue is just such a change of thinghood, not an alteration but the completion of the coming-into-being of a human being, just as putting on the roof completes the coming-into-being of a house (246a 17–246b 3). For a different reason, learning is not an alteration of the learner; knowing is a being-at-work that is always going on in us, unnoticed until we settle into it out of distraction and disorder (247b 17–18).

ambiguity (ὁμωνυμία *homōnymia*) The presence of more than one meaning in a word, sometimes by chance (as in "bark"), but more often by analogy or by derivation from one primary meaning. (See especially *Metaphysics* Book IV, Ch. 2.) A city or society is called healthy by analogy to an animal, a diet by derivation. Derived meanings may have many kinds of relation to the primary meaning, but all point to one thing (*pros hen*). Arrays of this truthful kind of ambiguity reflect causal structures in the world. Book V of the *Metaphysics*, mistakenly called a dictionary, is called by Aristotle the book about things meant in more than one way. Thomas Aquinas uses the word analogy to cover all non-chance ambiguity, but it makes a great difference to Aristotle that the meanings of *good* are unified only by analogy, while those of *being* point to one primary instance.

art (τέχνη *technē*) The know-how that permits any kind of skilled making, as by a carpenter or sculptor, or producing, as by a doctor or legislator. The artisan is not "creative"; in nature the form of the thing that comes into being is at work upon it directly, while in art the form is at work upon the soul of the artisan (1032b 14–15). Aristotle agrees with sculptors that Hermes is in the marble, and let out by taking away what obscures his image. Aristotle concludes that the origin of motion that produces statues is the art of sculpture, and incidentally the particular sculptor (195a 3–8). The artwork or artifact has no material cause proper to itself (192b 18–19—though a saw needs to be of a certain kind of material to hold an edge); in general the artisan uses the potencies of natural materials to counteract one another. The surface of a table strains to fall to earth, but the legs prevent it, while the legs strain to fall over and the tabletop prevents it, and similarly with the roof and walls of a house.

articulation (λόγος *logos*) The gathering in speech of the intelligible structure of anything, a combination of analysis and synthesis. A definition is one kind of articulation, but there are many others, including a ratio, a pattern, or reason itself. It can refer to anything that can be put into words—an argument, an account, a discourse, a story—or to the words into which anything is put—a word, a sentence, a chapter, a book. Translating *logos* as formula is misleading, since it has no implication of being the briefest, or any rigid, formulation of anything. In some translations, the word formula becomes a formula for a rich and varied idea; the word articulation is a slight improvement, used here wherever nothing better was appropriate.

being-at-work (ἐνέργεια *energeia*) An ultimate idea, not definable by anything deeper or clearer, but grasped directly from examples, at a glance or by analogy (1048 a 38–39). Activity comes to sight first as motion, but Aristotle's central thought is that all being is being-at-work, and that anything inert would cease to be. The primary sense of the word belongs to activities that are not motions; examples of these are seeing, knowing, and happiness, each understood as an ongoing state that is complete at every instant, but the human being that can experience them is similarly a being-at-work, constituted by metabolism. Since the end and completion of any genuine being is its being-at-work, the meaning of the word converges (1047a 30–31, 1050a 22–24) with that of the following:

being-at-work-staying-itself (ἐντελέχεια *entelecheia*) A fusion of the idea of completeness with that of continuity or persistence. Aristotle invents the word by combining ἐντελές *enteles* (complete, full-grown) with ἔχειν *echein* (= ἕξις *hexis*, to be a certain way by the continuing effort of holding on in that condition), while at the same time punning on ἐνδελέχεια *endelecheia* (persistence) by inserting τέλος *telos* (completion). This is a three-ring circus of a word, at the heart of everything in Aristotle's thinking, including the definition of motion. Its power to carry meaning depends on the working together of all the things Aristotle has packed into it. Some commentators explain it as meaning being-at-an-end, which misses the point entirely, and it is usually

translated as "actuality," a word that refers to anything, however triv-
ial, incidental, transient, or static, that happens to be the case, so that
everything is lost in translation, just at the spot where understanding
could begin.

cause (αἰτία *aitia*) The source of responsibility for anything. It thus
differs in two ways from its prevalent current sense: in always being a
source (1013a 17), rather the nearest agent or instrument that leads to
a result, and in referring more to responsibility for a thing's being as it
is than for its doing what it does. To understand anything is to know
its cause, and such an understanding is always incomplete without
an account of all four kinds of responsibility: as material, as form, as
origin of motion, and as end or completion (*Physics* Bk. II, Ch. 3).

chance (αὐτόματον *automaton* or τύχη *tuchē*) Any incidental cause.
At 197b 29–30, Aristotle invents the etymology τὸ αὐτό μάτην *to
auto matēn*, that which is itself in vain (but produces some other
result). Chance events or products always come from the interference
of two or more lines of causes; those prior causes always tend toward
natural ends or human purposes. Chance is thus derivative from the
"teleological" structure of the world, and is the reason nature acts for
the most part, rather than always, in the same way. In the *Physics*,
chance that is peculiarly relevant to a human being is distinguished as
fortune or luck under the name *tuchē*, but in the *Metaphysics* the two
senses are merged.

contradictory (ἀντίφασις *antiphasis*) One of a pair of opposites which
can have nothing between them, such as white and not-white.

contrary (ἐναντίον *enantion*) One of a pair of opposites which can have
something between them, such as white and black; but the opposition
need not be extreme, and could be between two shades of gray.

contemplation (θεωρία *theōria*) The being-at-work of the intellect (νοῦς
nous), a thinking that is like seeing, complete at every instant. Our
ordinary step-by-step thinking (διάνοια *dianoia*) aims at a completion
in contemplation, but it also presupposes an implicit contemplative
activity that is always present in us unnoticed. To know is not to achieve
something new, but to calm down out of the distractions of our native
disorder, and settle into the contemplative relation to things that is
already ours (247b 17–18). An analogy to the relation between step-
by-step reasoning and contemplation may be found in two ways of

looking at a painting or a natural scene; one's eyes may first roam from part to part, making connections, but one may also take in the sight whole, drinking it in with the eyes. The intellect similarly becomes most active when it comes to rest.

deprivation (στέρησις *sterēsis*) The absence in something of anything it might naturally have. Aristotle regards the distinction between the deprivation, which is opposite to form, and the material, which underlies and tends toward form, as a clarification of and advance over the opinions that came out of Plato's Academy (*Physics* Bk. I, Ch. 9).

end (τέλος *telos*) The completion toward which anything tends, and for the sake of which it acts. In deliberate action it has the character of purpose, but in natural activity it refers to wholeness. Aristotle does not say that animals, plants, and the cosmos *have* purposes but that they *are* purposes, ends-in-themselves. Whether any of them is in another sense for the sake of anything outside itself is always treated as problematic in the theoretical works (*Physics* 194a 34–36, *Metaphysics* 1072b 1–3, *On the Soul* 415b 2–3), though *Politics* 1257a 15–22 treats all other species as being for the sake of humans. As a settled opinion found throughout his writings, Aristotle's "teleology" is nothing but his claim that all natural beings are self-maintaining wholes.

example (ἐπαγωγή *epagōgē*) The perceptible particular, in which the intelligible universal is always evident. The word induction, which refers to a generalization from many examples, does not catch Aristotle's meaning, which is a "being brought face-to-face with" the universal present in each single example. A famous simile in the last chapter of the *Posterior Analytics* (100a 12–13) is often taken to mean that the universal must be built up out of particulars, just as a new position of a routed army is built up when many men have taken stands, but it means just the opposite: it only takes one man to take a stand, after which every other soldier, down to the original coward, will be identical to him. The rout corresponds to the condition of someone who has not yet experienced some universal in any of its instances. Evidence for this interpretation is found in many places, such as *Posterior Analytics* 71a 7–9 and *Physics* 247b 5–7, in which Aristotle unmistakably says that one particular is sufficient to make the universal known. That in turn is because the same form that is at work holding together the perceived thing is also at work on the soul of the perceiver (*On the Soul* 424a 18–19).

first philosophy (πρώτη φιλοσοφία *prōtē philosophia*) The study of immovable being, or of the sources and causes of all being. Aristotle's organized collection of writings on this topic was called by librarians τὰ μετὰ τὰ φυσικά *ta meta ta phusika*, "what comes after the study of natural things," but neither this phrase nor any like it is ever used by Aristotle. He names the topic in the order of the things themselves, rather than from the way we approach and think about them, and calls physics second philosophy. What we call metaphysics, the post-natural, is for Aristotle the pre-natural, the source and foundation of motion and change. That form is present in all things is a starting-point for physics; what form is must be clarified by first philosophy (192a 34–36).

form (μορφή *morphē* or εἶδος *eidos* or ἰδέα *idea*) Being-at-work (1050b 1–2). It is often said that Aristotle imports the form/material distinction from the realm of art and imposes it upon nature. In fact it is deduced in *Physics* Bk. I, Ch. 7 as the necessary condition of any change or becoming. In a compressed way in *Physics* Bk. II, Ch. 1, and more fully in *Metaphysics* Bk. VIII, Ch. 2, it is argued that arrangement is insufficient to account for form, which is evident only in the being-at-work of a thing. *Morphē* never means mere shape, but shapeliness, which implies the act of shaping, and *eidos,* after Plato has molded its use, is never the mere look of a thing, but its invisible look, seen only in speech (193a 31). *Idea,* from the same root as *eidos,* is used primarily when technical discussions within Plato's Academy are referred to, but the English words "idea" and "ideal" are distortions of it, suggesting something that can only be present in thought, which no one who used the Greek word intended.

genus (γένος *genos*) A divisible kind or class. It might arise from arbitrary acts of classification, in contrast to the *eidos* or species, the kind that exactly corresponds to the form that makes a thing just what it is. The highest general classes are the so-called "categories," the irreducibly many ways of attributing being. *Metaphysics* Bk. V, Ch. 7 lists eight of these: what something is, of what sort it is, how much it is, to what it is related, what it does, what is done to it, where it is, and when it is. *Categories* IV adds two more: in what position it is, and in what condition.

impasse (ἀπορία *aporia*) A logical stalemate that seems to make a question unanswerable. In fact, it is the impasses that reveal what the genuine questions are. Zeno's paradoxes are spectacular examples, resolved by Aristotle's definition of motion. In *Metaphysics* Book III, a collection of impasses in first philosophy, Aristotle writes "those who inquire without first being at an impasse are like people who do not know which way they need to walk" (995a 36–38). The word is often translated as "difficulty" or "perplexity," which are much too weak; it is only the inability to get past an impasse with one's initial presuppositions that forces the revision of a whole way of looking at things.

incidental (κατὰ συμβεβηκός *kata sumbebēkos*) Belonging to or happening to a thing not as a consequence of what it is. The word "accidental" is appropriate to some, but not all, incidental things; it is not accidental that the housebuilder is a flute player, but it is incidental. To any thing, an infinity of incidental attributes belongs, and this opens the door to chance (196b 23–29).

lead back (ἀνάγειν *anagein*) To produce an explanation while leaving the thing explained intact. Aristotle leads back all motion to change of place without reducing all motion to change of place.

material (ὕλη *hulē*) That which underlies the form of any particular thing. Unlike what we mean by "matter," material has no properties of its own, but is only a potency straining toward some form (192a 18–19). Bricks and lumber are material for a house, but have identities only because they are also forms for earth and water. The simplest bodies must have an underlying material that is not bodily (214a 13–16).

motion (κίνησις *kinēsis*) The being-at-work-staying-itself of a potency *as* a potency (*Physics* Bk. III, Chs. 1–3). Any thing is the being-at-work-staying-itself of a potency as material for that thing, but so long as that potency is at-work-staying-itself as a *potency*, there is motion (1048b 8–9). Motion is coextensive with, but not synonymous with change (μεταβολή *metabolē*). It has four irreducible kinds, with respect to thinghood, quality, quantity, and place. The last named is the primary kind of motion but involves the least change, so that the list is in ascending order of motions but descending order of changes.

mover (κινοῦν *kinoun*) Whatever causes motion in something else. The phrase "efficient cause" is nowhere in Aristotle's writings, and is highly misleading; it implies that the cause of every motion is a push or a pull. In *Physics* Bk. VII, Ch. 2, it is argued that in one way all motions lead back to pushes or pulls, but this is only a step in a long argument that concludes that every motion depends on a first mover that is motionless (258b 4–5), and the only kind of external mover that is included among the four kinds of cause in *Physics* Bk. II, Ch. 3 is the first origin of motion (194b 29–30). That there should be incidental, intermediate links by which motions are passed along when things bump explains nothing. That motion should originate in something motionless is only puzzling if one assumes that what is motionless must be inert; the motionless sources of motion to which Aristotle refers are fully at-work, and in their activity there is no motion because their being-at-work is complete at every instant (257b 9).

nature (φύσις *phusis*) The internal activity that makes anything what it is. The ideas of birth and growth, buried in the Latin origins of our word, are close to the surface of the Greek word, sprouting into all its uses. Nature is evident primarily in living things, but is present in everything nonliving as well, since it all participates in the single organized whole of the cosmos (1040b 5–10). Everything there is comes from nature, since all chance events and products result from the incidental interaction of two or more prior lines of causes, stemming from the goal-seeking activities of natural beings, and all artful making by human beings must borrow its material from natural things.

now (νῦν *nun*) The indivisible limit of a time. The word "moment" is not a suitable translation, because it refers to a *stretch* of a continuous process, nor is the word "instant" appropriate, since the now is relative to a soul that can recognize it. Time arises from the measurement of motion, which can only take place in a soul that can relate two motions by linking them to a now (223a 21–26).

number (ἀριθμός *arithmos*) Any multitude, whether of perceptible things, definite intelligible things, or empty units. The last named, the pure numbers of mathematics, Aristotle calls the numbers *by* which we count (219b 8), but the word normally refers to the first kind, the numbers which we count, such as the dozen eggs in a carton, a multitude *of* something. The remaining kind of number is alluded to at 206b 30–33; Plato seems to have taught that higher and lower forms are not related as genus and species but in the same way as a number and its units. That is, the unity of wisdom, courage, temperance, and justice, for example, would not be a common element contained in them all, but the sum of them all as virtue, the eidetic number four. A number in any of its senses is something discrete and countable, and never includes continuous magnitude; it therefore excludes fractions, irrationals, negatives, and all the other things brought under the idea of number when Descartes fused the ideas of multitude and magnitude, or of how-many and how-much, into one. It is thus a paradox, lost on us, when Aristotle says that time is both continuous and a number, but only in resolving that paradox is it possible to see how he understands time.

of-this-kind (ποιόν *poion*) Being of one or another sort is a more direct and immediate feature of things than having a quality (ποιότης *poiotēs*), a word that Aristotle rarely uses.

place (τόπος *topos*) The stable surroundings in which certain kinds of beings can sustain themselves, in which alone they can be at rest and fully active. When *dis*placed, anything strives to regain its appropriate place. This idea of place depends on the prior idea of the cosmos as an organized whole, in which there is no void. The contrary idea of space, as empty, homogeneous, and infinite, Aristotle regards as an abuse of mathematical abstraction: the positing of an extension of body without body.

potency (δύναμις *dunamis*) The innate tendency of anything to be at work in ways characteristic of the kind of thing it is; the way of being that belongs to material (1050a 18). The word has a secondary sense of mere logical possibility, applying to whatever admits of being true (1019b 31–33), but this is never the way Aristotle uses it. A potency in its proper sense will always emerge into activity, when the proper conditions are present and nothing prevents it (1047b 35–1048a 21).

primary (πρώτη, πρῶτον *protē, proton*) First in responsibility. It is translated as first when it means first in time.

rest (ἠρεμία *ēremia* or στάσις *stasis*) Motionlessness in whatever is naturally capable of motion (202a 4–5). A natural being at rest is still active. Nothing is inert.

sense perception (αἴσθησις *aisthesis*) Always the reception of organized wholes. Never sensation as meant by Hume or Kant, as the reception of isolated sense data. The primary object of sense perception is a *this*, a ready-made whole at which one may point.

separate (χωριστόν *chōriston*) Able to hang together as a whole, intact, on its own. Aristotle never uses the word to mean "separable." Mathematical things are not separate, not because they happen never to be found in isolation, but because they do not compose anything that could be at work. By the same token, the form *is* separate (1017b 27–28 and 1029a 27–30). When the form is remembered or reconstructed in thought as a universal, it is separate only in speech or articulation, but the form as it is in itself, as a being-at-work and a cause of being, is separate simply (1042a 32–33). In a number of places, such as 193b 4–5, Aristotle says that form is not separate except in speech, but this is always a first dialectical step, articulating the way form first comes to sight; at 194b 9–15 he already balances it with the opposite opinion, and points to the inquiry in which the question is resolved.

so much (ποσόν *poson*) Not isolated quantity but the muchness or manyness that belongs to something. The former is studied by the mathematician; the latter is present in nature.

source (ἀρχή *archē*) A ruling beginning. It can refer to the starting point of reasoning, but the usual translations "principle" or "first principle" are rarely adequate, since the word most often refers to a being rather than to a proposition or rule. The divine intellect deduced in Bk. XII of the *Metaphysics* is not an explanatory principle but a being on which all other beings depend.

thinghood (οὐσία *ousia*) The way of being that belongs to anything which has attributes but is not an attribute of anything, which is also separate and a *this* (1028b 38–39, 1029a 27–28). Whatever has being in this way is an independent thing. In ordinary speech the word means wealth or inalienable property, the inherited estate that cannot be taken away from one who is born with it. Punning on its connection with the participle of the verb "to be," Plato appropriates the word (as at *Meno* 72B) to mean the very being of something, in respect to which all instances of it are exactly alike. Aristotle elaborates this meaning into a distinction between the thinghood of a thing and the array of attributes—qualities, quantities, relations, places, times, actions, and ways of being acted upon—that can belong to it fleetingly, incidentally, derivatively, and in common with things of other kinds. He concludes that thinghood is not reducible to any sum of attributes (1038b 25–28, 1039a 1–3). It thus denotes a fullness of being and self-sufficiency which the Christian thinker Augustine did not believe could be present in a created thing (*City of God* XII, 2); he concluded that, while *ousia* meant *essentia*, things in the world possess only deficient kinds of being. *Substantia*, the capacity to have predicates, became the standard word in the subsequent Latin tradition for the being of things. A blind persistence in this tradition gave us "substance" as the translation of a word that it was conceived as negating.

this (τόδε τι *tode ti*) That which comes forth to meet perception as a ready-made, independent whole. A *this* is something that can be pointed at, because it holds together as separate from its surroundings, and need not be constructed or construed out of constituent data, but stands out from a background. The mistranslation "this somewhat" reads the phrase backward, and is flatly ruled out by many passages, such as 1038b 25–28.

underlying thing (ὑποκείμενον *hupokeimenon*) That in which anything inheres. It can be of various kinds. Change presupposes something that persists. Attributes belong to some whole that it not just their sum. Form works on some material. An independent thing is an underlying thing in the first two ways, but not in the third (1029a 20–28).

universal (καθόλου *katholou*) Any general idea, common property, or one-applied-to-many. It is never separate and can have no causal responsibility, unlike the form, which is a being-at-work present in things, making them what they are (1040b 28–30, 1041a 4–5).

virtue (ἀρετή *aretē*) Any of the excellences of the human soul, primarily wisdom, courage, moderation, and justice. Though they depend on learning or habituation, Aristotle regards them as belonging to our nature. Without them we are like houses without roofs, not fully what we are (246a 17–246 b3).

what it is for something to be (τί ἦν εἶναι *ti ēn einai*) What anything keeps on being, in order to be at all. The phrase expands τί ἐστι *ti esti*, what something is, the generalized answer to the question Socrates asks about anything important: "What is it?" Aristotle replaces the bare "is" with a progressive form (in the past, but with no temporal sense, since only in the past tense can the progressive aspect be made unambiguous) plus an infinitive of purpose. The progressive signifies the continuity of being-at-work, while the infinitive signifies the being-something or independence that is thereby achieved. The progressive rules out what is transitory in a thing, and therefore not necessary to it; the infinitive rules out what is partial or universal in a thing, and therefore not sufficient to make it be. The learnèd word "essence" contains nothing of Aristotle's simplicity or power.

Aristotle's
Metaphysics

Book I (Book A)
Wisdom[1]

Chapter 1 All human beings by nature stretch themselves out toward knowing. A sign of this is our love of the senses; for even apart from their use, they are loved on their own account, and above all the rest, the one through the eyes. For not only in order that we might act, but even when we are not going to act at all, we prefer seeing, one might say, as against everything else. And the cause is that, among the senses, this one most of all makes us discover things, and makes evident many differences. By nature, then, the animals come into being having sense perception, though in some of them memory does not emerge out of this, while in others it does. And for this reason, these latter are more intelligent and more able to learn than those that are unable to remember, while as many of them as are not able to hear sounds are intelligent without learning (such as a bee, or any other kind of animal that might be of this sort), but as many do learn as have this sense in addition to memory. So the other animals live by images and memories, but have a small share of experience, but the human race lives also by art and reasoning. And for human beings, experience arises from memory, since many memories of the same thing bring to completion a capacity for one experience.

Now experience seems to be almost the same thing as knowledge or art, but for human beings, knowledge and art result from experience, for experience makes art, as Polus says and says rightly, but inexperience makes chance. And art comes into being whenever, out of many conceptions from experience, one universal judgment arises about those that are similar. For to have a judgment that this thing was beneficial to Callias when he was sick with this disease, and to Socrates, and one by one in this way to many people, belongs to experience. But the judgment that it was beneficial to all such people, marked out as being of one kind, when they were sick with this disease, such as to sluggish or irritable people[2] when they were feverish

---

[1] This title for Book I supplied by the translator.

[2] The Greek words imply the predominance, respectively, of phlegm or yellow bile, two of the four humors whose imbalance was thought to produce many diseases.

with heat, belongs to art. For the purpose of acting, experience doesn't seem to differ from art at all, and we even see people with experience being more successful than those who have a rational account without experience. (The cause of this is that experience is familiarity with things that are particular, but art with those that are universal, while actions and all becoming are concerned with what is particular. For the doctor does not cure a human being, except incidentally, but Callias or

981a 20   Socrates or any of the others called by such a name, who happens to be a human being. So if someone without experience has the reasoned account and is familiar with the universal, but is ignorant of what is particular within it, he will often go astray in his treatment, since what is treated is particular.)

Nevertheless, we think that knowing and understanding are present in art more than is experience and we take the possessors of arts to be wiser than people with experience, as though in every instance wisdom is more something resulting from and following along with knowing; and this is because the ones know the cause while the others do not. For people with experience know the what, but do not

981a 30   know the why, but the others are acquainted with the why and the cause. For this reason we also think the master craftsmen in each kind

981b   of work are more honorable and know more than the manual laborers, and are also wiser, because they know the causes of the things they do,[3] as though people are wiser not as a result of being skilled at action, but as a result of themselves having the reasoned account and knowing the causes. And in general, a sign of the one who knows and the one who does not is being able to teach, and for this reason we regard the art, more than the experience, to be knowledge, since the ones can, but the others cannot, teach.

981b 10   Further, we consider none of the senses to be wisdom, even though they are the most authoritative ways of knowing particulars; but they do not pick out the why of anything, such as why fire is hot, but only that it is hot. So it is likely that the one who first discovered any art whatever that was beyond the common perceptions was wondered at by people, not only on account of there being something useful in

---

[3] Some of the manuscripts have the following insertion here: "The others, as do also some of the things without souls, do what they do without knowing, as fire burns, the soulless things doing each of these things by some nature, but the manual laborers by habit."

his discoveries, but as someone wise and distinguished from other people. But once more arts had been discovered, and some of them were directed toward necessities but others toward a way of living, it is likely that such people as were discoverers of the latter kind were always considered wiser, because their knowledge was not directed toward use. Hence when all such arts had been built up, those among the kinds of knowledge directed at neither pleasure nor necessity were discovered, and first in those places where there was leisure. It is for this reason that the mathematical arts were first constructed in the neighborhood of Egypt, for there the tribe of priests was allowed to live in leisure.

981b 20

Now it has been said in the writings on ethics what the difference is among art, demonstrative knowledge, and the other things of a similar kind, but the purpose for which we are now making this argument is that all people assume that what is called wisdom is concerned with first causes and origins. Therefore, as was said above, the person with experience seems wiser than those who have any perception whatever, the artisan wiser than those with experience, the master craftsman wiser than the manual laborer, and the contemplative arts more so than the productive ones. It is apparent, then, that wisdom is a knowledge concerned with certain sources and causes.

981b 30

982a

Chapter 2 Since we are seeking this knowledge, this should be examined: about what sort of causes and what sort of sources wisdom is the knowledge. Now if one takes the accepted opinions we have about the wise man, perhaps from this it will become more clear. We assume first that the wise man knows all things, in the way that it is possible, though he does not have knowledge of them as particulars. Next, we assume that the one who is able to know things that are difficult, and not easy for a human being to know, is wise; for perceiving is common to everyone, for which reason it is an easy thing and nothing wise. Further, we assume the one who has more precision and is more able to teach the causes is wiser concerning each kind of knowledge. And among the kinds of knowledge, we assume the one that is for its own sake and chosen for the sake of knowing more to be wisdom than the one chosen for the sake of results, and that the more ruling one is wisdom more so than the more subordinate one; for the wise man ought not to be commanded but to give orders, and ought not to obey someone else, but the less wise ought to obey him.

982a 10

982a 20   We have, then, such and so many accepted opinions about wisdom and those who are wise. Now of these, the knowing of all things must belong to the one who has most of all the universal knowledge, since he knows in a certain way all the things that come under it; and these are just about the most difficult things for human beings to know, those that are most universal, since they are farthest away from the senses. And the most precise of the kinds of knowledge are the ones that are most directed at first things, since those that reason from fewer things are more precise than those that reason from extra ones, as arithmetic is more precise than geometry. But surely the skill that is suited to teach is the one that has more insight into causes, for those

982a 30   people teach who give an account of the causes about each thing. And knowing and understanding for their own sakes belong most to the knowledge of what is most knowable. For the one who chooses what

982b   is known through itself would most of all be choosing that which is knowledge most of all, and of this sort is the knowledge of what is most knowable. But what are most knowable are the first things and the causes, for through these and from these the other things are known, but these are not known through what comes under them. And the most ruling of the kinds of knowledge, or the one more ruling than what is subordinate to it, is the one that knows for what purpose each thing must be done; and this is the good of each thing, and in general the best thing in the whole of nature. So from all the things that have been said, the name sought falls to the same kind of knowledge, for it

982b 10   must be a contemplation of the first sources and causes, since also the good, or that for the sake of which, is one of the causes.

That it is not a productive knowledge is clear too from those who first engaged in philosophy. For by way of wondering, people both now and at first began to philosophize, wondering first about the strange things near at hand, then going forward little by little in this way and coming to impasses about greater things, such as about the attributes of the moon and things pertaining to the sun and the stars and the coming into being of the whole. But someone who wonders and is at an impasse considers himself to be ignorant (for which reason the lover of myth is in a certain way philosophic, since a myth is composed

982b 20   of wonders). So if it was by fleeing ignorance that they philosophized, it is clear that by means of knowing they were in pursuit of knowing, and not for the sake of any kind of use. And the following testifies to the same thing: for it was when just about all the necessities were present, as well as things directed toward the greatest ease and recreation, that this kind of understanding began to be sought. It is clear then that we seek it for no other use at all, but just as that human being is free, we say, who has his being for his own sake and not for the sake of someone else, so also do we seek it as being the only one of the kinds of knowledge that is free, since it alone is for its own sake.

For this reason one might justly regard the possession of it as not appropriate to humans. For in many ways human nature is slavish, so that, according to Simonides, "only a god should have this honor," but a man is not worthy of seeking anything but the kind of knowledge that fits him. If indeed the poets have a point and it is the nature of the divine power to be jealous, it would be likely to happen most of all in this case, and all extraordinary people would be ill-fated. But it is not even possible for the divine power to be jealous, but according to the common saying "many lyrics are lies," and one ought not to regard anything else as more honorable than this knowledge. For the most divine is also the most honorable, and this knowledge by itself would be most divine in two ways. For what most of all a god would have is that among the kinds of knowledge that is divine, if in fact any of them were about divine things. But this one alone happens to have both these characteristics; for the divine seems to be among the causes for all things, and to be a certain source, and such knowledge a god alone, or most of all, would have. All kinds of knowledge, then, are more necessary than this one, but none is better.

It is necessary, however, for the possession of it to settle for us in a certain way[4] into the opposite of the strivings with which it began. For everyone begins, as we are saying, from wondering whether things are as they seem, such as the self-moving marvels,[5] or about the reversals of the sun[6] or the incommensurability of the diagonal (for it seems

982b 30

983a

983a 10

---

[4] Contemplation and wonder are opposite only in a certain way; they are alike in being active, humble, and appreciative. There is all the difference in the world between problem solving, which sets aside a solved problem like a finished crossword puzzle, and an inquiry that works through wonder and out the other side.

[5] An early commentator describes these as toys displayed in magic shows, and *Mechanics* 848a 20–38 describes a way that mechanical marvels were moved by concealed gears.

[6] If one pays attention to the ascent and descent of the sun in the sky from day to day, the solstices are astounding events. The sun stands still at the same height for a few days before reversing direction.

amazing to all those who have not yet seen the cause if anything
is not measured by the smallest part).7 But it is necessary to end in
what is opposite and better, as the saying goes, just as in these cases

983a 20   when people understand them; for nothing would be so surprising to
a geometer as if the diagonal were to become commensurable. What,
then, is the nature of the knowledge being sought, has been said, and
what the object is on which the inquiry and the whole pursuit must
alight.

Chapter 3 Since it is clear that one must take hold of a knowledge
of the causes that originate things (since that is when we say we know
each thing, when we think we know its first cause), while the causes
are meant in four ways, of which one is thinghood, or what it is for
something to be (since the why leads back to the ultimate reasoned

983a 30   account, and the first why is a cause and source), another is the material
or underlying thing, a third is that from which the source of motion
is, and the fourth is the cause opposite to that one, that for the sake
of which or the good (since it is the completion of every coming-into-
being and motion), which have been sufficiently looked into by us

983b   in the writings about nature, still, let us take up also those who came
before us into the inquiry about beings and philosophized about truth.
For it is clear that they too speak of certain sources and causes. So for
those who go back over these things, there will be some profit for the
present pursuit; for we will either find out some other kind of cause
or be more persuaded about the ones we are now speaking of.

Of those who first engaged in philosophy, most thought that the
only sources of all things were of the species of material; that of which
all things are made, out of which they first come into being and into

983b 10   which they are at last destroyed, its thinghood abiding but changing
in its attributes, this they claim is the element and origin of things,
for which reason nothing ever comes into being or perishes, since this
sort of nature is always preserved, just as we would not say either
that Socrates simply comes into being when he becomes beautiful or
educated in refined pursuits, or that he perishes when he sheds these
conditions, since the underlying thing, Socrates himself, persists, so

---

7 The Pythagorean demonstration that no fraction of the side of a square, however
small, could fit any exact number of times into its diagonal, is given in the Dover
edition of Euclid's *Elements*, Vol. III, p. 2.

neither does anything else. For there must be some nature, either one or multiple, out of which the other things come into being while that one is preserved. About the number and kind of such sources, however, they do not all say the same thing, but Thales, the founder of this sort of philosophy, says it is water (for which reason too he declared that the earth is on water), getting hold of this opinion perhaps from seeing that the nourishment of all things is fluid, and that heat itself comes about from it and lives by means of it (and that out of which things come into being is the source of them all). So he got hold of this opinion by this means, and because the seeds of all things have a fluid nature, while water is in turn the source of the nature of fluid things.

983b 20

There are some who think that very ancient thinkers, long before the present age, who gave the first accounts of the gods, had an opinion of this sort about nature. For they made Ocean and Tethys the parents of what comes into being, and made the oath of the gods be by water, called Styx by them; for what is oldest is most honored, and that by which one swears is the most honored thing. But whether this opinion about nature is something archaic and ancient might perhaps be unclear, but Thales at least is said to have spoken in this way about the first cause. (One would not consider Hippo worthy to place among these, on account of his cut-rate thinking.) Anaximenes and Diogenes set down air as more primary than water and as the most originative of the simple bodies,[8] while Hippasus of Metapontium and Heracleitus of Ephesus set down fire, and Empedocles, adding earth as a fourth to those mentioned, sets down the four (for he says these always remain and do not come into being except in abundance or fewness, being combined and separated into or out of one). Anaxagoras of Clazomenae, who was before Empedocles in age but after him in his works, said the sources were infinite; for he said that almost all homogeneous things are just like water or fire in coming into being or perishing only by combination and separation, but otherwise neither come into being nor perish but remain everlasting.

983b 30

984a

984a 10

From these things, then, one might suppose that the only cause is the one accounted for in the species of material; but as people went forward in this way, their object of concern itself opened a road for

---

[8] Aristotle does not consider earth, air, fire, and water elements, since they can turn into each other, but only as the simplest bodies. What underlies them is no longer bodily.

984a 20   them, and contributed to forcing them to inquire along it. For no matter how much every coming-into-being and destruction is out of some one or more kinds of material, why does this happen and what is its cause? For surely the underlying material itself does not make itself change. I mean, for example, neither wood nor bronze is responsible, respectively, for its own changing, nor does the wood make a bed or the bronze a statue, but something else is responsible for the change. But to inquire after this is to seek that other kind of source, which we would call that from which the origin of motion is. Now some of these who from the very beginning applied themselves to this sort of pursuit and said that the underlying material was one thing, were not at all

984a 30   displeased with their own accounts, but some of those who said it was one, as though defeated by this inquiry, said that the one and the whole of nature were motionless, not only with respect to coming into being and destruction (for this is present from the beginning and everyone

984b   agrees to it), but also with respect to every other kind of change, and this is peculiar to them. So of those who said that the whole is one, none happened to catch sight of this sort of cause unless in fact Parmenides did, and he only to the extent that he set down the causes as being not only one but in some way two.[9] But it is possible to say so about those who made things more than one, such as hot and cold, or fire and earth; for they use the nature of fire as having the power to set things in motion, but water and earth and such things as the opposite.

But after these people and sources of this kind, since they were not

984b 10   sufficient to generate the nature of things, again by the truth itself, as we say, people were forced to look for the next kind of source. For that some beings are in a good or beautiful condition, or come into being well or beautifully, it is perhaps not likely that fire or earth or any other such thing is responsible, nor that they would have thought so. Nor, in turn, would it be a good idea to turn over so great a concern to chance or luck. So when someone said an intellect was present, just as in animals, also in nature as the cause of the cosmos and of all order, he looked

---

[9] Parmenides's poem can be read as a complex refutation of the sort of account given by such thinkers as Thales and Anaximenes, who said the world was one permanent material which turned into the many things of experience. On the basis of reason (Parmenides's way of truth), the one being could not change, while on the basis of appearance (his way of seeming), there must be two ultimate causes at work (light and night).

like a sober man next to people who had been speaking incoherently beforehand. Obviously we know that Anaxagoras reached as far as saying these things, but Hermotimus of Clazomenae is given credit for saying them earlier. Those, then, who took things up in this way set down a source which is at the same time the cause of the beautiful among things and the sort of cause from which motion belongs to things.

Chapter 4 One might suspect that Hesiod was the first one to seek out such a thing, or someone else who had set down love or desire among the beings as a source, as Parmenides also did; for he, in getting things ready for the coming into being of the whole, says that first "of all the gods, [the all-governing divinity] devised love," while Hesiod says

Chaos came into being as the very first of all things, but then
Broad-breasted earth....and also
Love, who shines out from among all the immortals,
as though there needed to be present among beings some sort of cause 984b 30 that would move things and draw them together. Now how I ought to distribute their portions to them about who was first, permit me to postpone judging. But since the opposites of the good things are obviously also present in nature, and there is not only order and beauty 985a but also disorder and ugliness, and more bad and ordinary things than good and beautiful ones, in this way someone else brought in friendship and strife, each as the cause of one of these kinds of thing. For if one were to pursue and get hold of Empedocles' thinking, rather than what he said inarticulately, one would find that friendship was the cause of the good things and strife of the bad. So if one were to claim that Empedocles both says and is the first to say that bad and good are sources, one would perhaps speak rightly, if in fact the cause of all good things is the good itself. 985a 10

So these people, as we are saying, evidently got this far with two causes out of those we distinguished in the writings about nature, the material and that from which the motion is, but did so dimly and with no clarity, rather in the way nonathletes do in fights; for while dancing around they often land good punches, but they do not do so out of knowledge, nor do these people seem to know what they are saying. For it is obvious that they use these causes scarcely ever, and only to a tiny extent. For Anaxagoras uses the intellect as

a makeshift contrivance for cosmos production, and whenever he
985a 20   comes to an impasse about why something is necessarily a certain
way, he drags it in, but in the other cases he assigns as the causes of
what happens everything but the intellect; and Empedocles, though
he uses his causes more than that, surely does not either use them
sufficiently or come up with any consistency with them. Certainly in
many places friendship separates things for him, and strife combines
things. For whenever the whole is divided by strife into its elements,
fire is combined into one, as is each of the other elements; but whenever
they come back together into one, the parts must be separated back out
of each element. Empedocles was the first who, beyond those before
985a 30   him, brought in this sort of cause by dividing it, making the source of
motion not one thing but different and opposite ones, and furthermore,
*   he first spoke of the so-called four elements as the causes in the species
of material. (In fact, though, he didn't use them as four, but as though
985b   they were only two: fire on one side by itself, and its opposites, earth,
air, and water, as one nature on the other side. One may get this by
looking carefully at what is in the verses.)

So as we are saying, he claimed that the sources were of this sort
and this many. But Leucippus and his colleague Democritus say the
elements are the full and the void, of which the one, as what is, is full
and solid, while the other, what is not, is void (for which reason they
say that being in no sense *is* more than nonbeing, nor body more so
985b 10   than void), and that these are responsible for things as material. And
just as those who make the underlying being one, and generate the
other things by means of modifications of it, set down the rare and
the dense as the sources of the modifications, in the same way these
people too say that the differences in the material are responsible for
the other things. They, however, say these differences are three: shape,
order, and position. For they say that what is differs only by means of
"design, grouping, and twist," but of these, design is shape, grouping
is order, and twist is position. For A differs from N in shape, AN from
NA in order, and Z from N in position. As for motion, from what source
985b 20   or in what way it belongs to things, these people, much like the others,
lazily let it go. So about the two causes, as we are saying, the inquiry
seems to have gone this far on the part of our predecessors.

Chapter 5 But among these people and before them, those who are
called Pythagoreans, taking up mathematical things, were the first to

promote these, and having been reared on them, they supposed that the sources of them were the sources of all things. And since numbers are by nature primary among these, and in them they thought they saw many similarities to the things that are and come to be, more so than in fire or earth or water—that such-and-such an attribute of numbers was justice, so-and-so is soul or intellect, another one due measure, 985b 30 and likewise with, one might say, everything—and what's more, saw in numbers the properties and explanations of musical harmonies, since then the entire nature of the other things seemed to be after the likeness of numbers, and the numbers seemed to be the primary 986a things in all nature, they assumed that the elements of numbers were the elements of all things, and that the whole heaven was a harmony and a number. And as many things among numbers and harmonies as had analogies to the attributes and parts of the heavens and to the whole cosmic array, they collected and fit together. And if anything was left out anywhere, they were persistent, so that everything would be strung together for them as a system. I mean, for example, since the number ten seemed to be complete and to include the whole nature of numbers,[10] they said that the things that move through the heavens 986a 10 are also ten, but since there are only nine visible, for this reason they made the tenth the counter-earth. But this is set out more precisely by us in other writings.[11]

But the purpose for which we are recounting these things is this: that we might understand from these people both what they set down as causes, and how they fall in with the kinds of causes described. Now it is obvious that they consider number to be a cause both as material for things and as responsible for their attributes and states; that they consider the elements of number to be the even and the odd, of which

---

[10] Ten is not a "perfect" number in the sense that six is, as the sum of all its factors, but it is, for example, 1+2+3+4. Speusippus, Plato's nephew and successor in the Academy, was an ardent Pythagorean who devoted half a book to the properties of the number ten, as a form that included all the patterns (triangular, etc.) that are evident in other numbers.

[11] See *On the Heavens*, Bk. II, Ch. 13. The Pythagoreans, among other ancients, put the sun at the center and the earth in motion. The counter-earth was conceived as a dark planet, moving at the same rate as the earth, always concealed by the sun. (The number of nine celestial bodies counts the sphere of the fixed stars as one thing, along with the five visible planets and the earth, moon, and sun.)

986a 20

the latter is limited but the former unlimited[12]; that they consider the number one to be composed of both of these (for it is both even and odd[13]); and that they consider number to arise from the one, and the whole heaven, as was said, to be numbers.

Various ones of these same people say that there are ten causes, gathered into a series of corresponding pairs:

limit	unlimited
odd	even
one	many
right	left
male	female
still	moving
straight	crooked
light	dark
good	bad
square	oblong.

986a 30

986b

In this way too Alcmaeon of Croton seems to have taken it up, and he either got it from them or they got this account from him, since Alcmaeon expressed himself in much the way they did. For he says that many human things are twofold, referring not just to the oppositions as they distinguished them, but to any there might happen to be, such as white/black, sweet/bitter, good/bad, big/little. So he spread out indiscriminate statements about the rest of them, but the Pythagoreans declared how many and what the oppositions were. From both Alcmaeon and the Pythagoreans there is this much that one may gather, that the sources of things are contraries, and from the latter, how many and what these are. But in what way it is possible to bring them into line with the kinds of causes described was not clearly articulated by these people, though they seem to rank them as in the species of material, since they say that being is put together and molded out of these contraries as constituents.

986b 10

So of the ancients who said that the elements of nature are more than one, it is also sufficient from the preceding to see their thinking.

---

[12] The even breaks in half; the extra unit in the odd binds it into a whole.

[13] The odd numbers arise from the unit when it is added to the evens, but the even numbers also arise from the unit, since it must be doubled to produce the first of them. One itself is not properly a number at all, but that which generates all species of them, and is potentially each.

But there are some who declared about the whole that it is one nature, though not all in the same way, either in how good their thinking was or with respect to the nature. An account about them in no way fits into the present examination of the causes (for they do not speak in the same way as some writers about nature, who set down being as one but generate things out of the one as from material, but in a different way, since those others add in motion when they generate the whole, but these people say that it is motionless). Nonetheless, it can find a home in this inquiry to this extent: Parmenides seems to take hold of what is one according to reason, Melissus of what is one in material (on account of which one says that it is finite, the other infinite). But Xenophanes, the first of these who made things one (for Parmenides is said to have become a student of his), made nothing clear, nor does he seem to have made contact with nature in either of these two ways, but gazing off into the whole heaven, he said that divinity was one. So these, as we say, must be dismissed from the present inquiry, two of them completely as being a little too crude, Xenophanes and Melissus, though Parmenides to some degree seems to speak with more insight; for since he thought it fit that besides being, nonbeing could not be in any way, he necessarily supposed that being is one and that there is nothing else (about which we have spoken more clearly in the writings about nature[14]), but being forced to follow appearances, and assuming that what is one from the standpoint of reason is more than one from the standpoint of perception, he set down in turn two causes and two sources, a hot one and a cold one, as though speaking of something like fire and earth, and of these he ranks the hot one under being and the other under nonbeing.

986b 20

986b 30

987a

So from what has been said, and from the wise men who have so far sat in council in our account, we have ascertained these things: from the first ones that the origin of things is corporeal (since water and fire and things of that sort are bodies), and from some of them that the corporeal source is one, from others more than one, both kinds, however, placing them as in the species of material; and from others that there is this sort of cause and in addition to it one from which the motion is, while from some of them that this cause is one, from others

---

14 See *Physics*, Bk. I, chapters 2 and 3, especially 187a 3–11.

987a 10   that it is twofold. So up to the time of the Italians,[15] and apart from them, the others have spoken about these things in ways that made them murkier, except, as we say, they happened to use two kinds of cause, and one of these, that from which the motion is, some made one and others twofold. Now the Pythagoreans have said in the same way that the sources are two, but they added on top of that this much that is peculiar to them, that they did not consider the limited and the unlimited to be natures belonging to any other things, such as fire or earth or anything else of that sort, but that the unlimited itself and the one itself are the thinghood of the things to which they are attributed,

987a 20   for which reason also, number is the thinghood of all things. About these things, then, they spoke in this way, and they were the first to speak about and define the what-it-is of things, though they handled it in too simple a way. For they defined superficially, and the primary thing to which the stated definition belonged, they considered to be the thinghood of the thing, just as if one were to suppose that the number two and double were the same thing, because doubleness belongs first of all to the number two. But still it is not the same thing to be double and to be two; otherwise what is one would be many, which in fact is the way it turned out for these people. So from those who came before us, there is this much to grasp.

Chapter 6 After the philosophic speculations that have been men-
987a 30   tioned came the careful work of Plato, which in many ways followed the lead of these people, but also had separate features that went be-yond the philosophy of the Italians. For having become acquainted from youth at first with Cratylus and the Heracleitean teachings that all sensible things are always in flux and that there is no knowledge
987b   of them, he also conceived these things that way later on. And since Socrates exerted himself about ethical matters and not at all about the whole of nature, but in the former sought the universal and was the first to be skilled at thinking about definitions, Plato, when he adopted this, took it up as applying to other things and not to sensible ones, because of this: it was impossible that there be any common definition of any of the perceptible things since they were always changing. So he called this other sort of beings forms, and said the perceptible things

---

[15] That is, the Pythagoreans.

were apart from these and all spoken of derivatively from these, for the
many things with the same names as the forms were results of partici-   987b 10
pation. He changed only the name participation, for the Pythagoreans
said that beings *are* by way of imitation of the numbers, but Plato by
way of participation, having changed the name. What this participation or imitation of the forms might be, however, they were in unison
in leaving behind to be sought.

What's more, apart from the sensible things and the forms, he said
there were among things mathematical ones, in between, differing
from the sensible ones in being everlasting and motionless, and from
the forms in that there are any multitude of them alike, while each form
itself is only one. And since the forms are the causes of the others, he
thought that the elements of the forms were the elements of all beings.   987b 20
As material, then, the great and the small were the sources, and as
thinghood, the one, for out of the former, by participation in the one,
the forms are composed as numbers.[16] So in saying that the one was an
independent thing, and not any other thing said to be one, he spoke in
much the same way as the Pythagoreans, and in saying that numbers
were the causes of thinghood for other things he spoke exactly as they
did; but to have made a dyad in place of the infinite as one thing, and
to have made the infinite out of the great and the small, was peculiar
to him. It was also peculiar to him to set the numbers apart from the
perceptible things, while they had said that the things themselves were
numbers, and they did not set the mathematical things between them.
Now his having made the one and the numbers be apart from the things   987b 30
we handle, and not the same way the Pythagoreans had said, and his
introduction of the forms came about because of his investigation in the

---

[16] The Greek word *arithmos* does not refer only to a mathematical number, but to a
multitude of any kind. Here it is applied to the forms as distinct from mathematical
things. The form is conceived not as a common element in things, but as itself
an assemblage of intelligible elements, having the unity conferred by the one, as
well as the internal diversity arising out of indefinite duality. This technical side of
Plato's teaching appears only in glimpses in the dialogues, where the one appears
as the beautiful itself in the *Symposium*, or as the good itself, beyond being, in the
Republic, and where the dyad lies behind the puzzle of the relation of motion, rest,
and being in the *Sophist*. Being is not a third thing that motion and rest share, but
is the determinate twofoldness of motion and rest themselves. See Jacob Klein, *Greek
Mathematical Thought and the Origin of Algebra* (M.I.T. Press, 1968), Ch. 7, and Robert
Williamson, "*Eidos* and *Agathon* in Plato's *Republic*," *The St. John's Review*, XXXIX, 1–2
(1989–90).

988a

realm of definitions (for the earlier thinkers had no part in dialectic), but his having made the other nature a dyad was so that the numbers, outside of the primes, might be generated out of it in a natural way, as though from some sort of modeling clay.

But surely things happen in the opposite way, for this way is not reasonable. For they make many things out of this material, while the form generates only once, but it is apparent that from one material comes one table, while the person who brings the form to bear, though he is one, makes many tables. And it is similar with the male in relation to the female; for she becomes pregnant by one act of intercourse, while the male is the cause of many pregnancies, and surely these things are images of the origins of things. But Plato made distinctions in this way about the things that are being sought, and it is clear from what has

988a 10

been said that he used only two causes, the one that is responsible for the what-it-is and the one that results from material (for the forms are causes for the other things of what they are, and the one is such a cause for the forms). And it is clear what the underlying material is said to be, over against which the forms are, in the case of perceptible things, and the one, among the forms, that it is the dyad of the great and the small. And further, he referred the cause of what is good and bad respectively to each of the two elements, just as we say certain of the earlier philosophers, such as Empedocles and Anaxagoras, were trying to find a way to do.

Chapter 7 In a curtailed way, then, and hitting the high spots, we

988a 20

have gotten hold of who happens to have spoken about origins and truth, and in what way. Still, we get this much from them: that of those who have spoken about origin and cause, not one has said anything that went outside those that were distinguished by us in the writings on nature, but all of them, though murkily, have obviously touched on them in some way. For some speak of the source as material, whether they set it down as one or more than one, and whether they set it up as a body or as bodiless. (For example, Plato speaks of the great and the small, the Italians of the unlimited, Empedocles of fire, earth, water, and air, and Anaxagoras of the infinity of homogeneous things; now all of these have gotten hold of this sort of cause, as also have all those

988a 30

who speak of air or fire or water or something denser than fire but less dense than air, since some people have said that the primary element is of this sort.)

So these people have grasped only this sort of cause, but some others have grasped the sort from which the source of motion is (as have all those who make friendship and strife, or intellect, or love a source). But about what it is for something to be, and thinghood, no one has delivered up a clear account, but those who posit the forms speak of it most. (For they do not take up the forms as material of perceptible things or the one as material of the forms, nor either the forms or the one as anything from which the source of motion comes—for they say rather that they are responsible for motionlessness and being at rest—but they offer the forms as what it is for each of the other things to be, and the one as what it is for the forms to be.)

988b

That for the sake of which actions and changes and motions are, they speak of as a cause in a certain way, but they do not say it that way, nor speak of what is so by its very nature. For those who speak of intellect or friendship as good set these up as causes, but do not speak as though anything that is either has its being or comes into being for the sake of these, but as though motions arose from these. And in the same way too, those who speak of the one or being as such a nature do say that it is the cause of thinghood, but not that it either is or comes about for the sake of this; so it turns out that they in a certain way both say and do not say that the good is a cause, since they say it is so not simply but incidentally.[17] So all these people seem also to bear witness that the right distinctions have been made by us about the causes, as to both how many and of what sort they are, since they have not been able to touch on another kind of cause; and it is clear in addition that the sources of things must be sought either all in this way or in some particular way from among them. After this, then, let us go through the possible impasses that concern the way each of these people spoke and how each one stands toward the sources.

988b 10

988b 20

Chapter 8 Now it is clear that all those who set down the whole as one, and some one nature as material, and this as bodily and having magnitude, erred in many ways. For they set down elements of bodies

---

17 Again, one must distinguish what is in Plato's dialogues from his unwritten technical doctrines. In the dialogues there is no "theory of forms," but only a number of dialectical invitations to the reader to think in such a direction. On the other hand, the classic statement in all literature that an adequate explanation of anything requires knowing the cause for the sake of which it is, may be found in the *Phaedo* at 97B–99D.

only, but not of bodiless things, though there are also bodiless ones. And when they turn their hands to giving an account of the causes of coming-into-being and destruction, in fact when they give accounts of the natures of all things, they abolish the cause of motion. What's more, they err by not setting down thinghood as a cause at all, nor the what-

988b 30

it-is of things, and on top of this by casually calling any one whatever of the simple bodies, except earth, a source, without examining the way they are made by coming into being out of one another; I am speaking of fire, water, earth, and air. For some of them do come into being out of others by combination, and others by separation, but this makes the greatest difference toward being more primary and more derivative. For in one sense the most elementary thing of all would seem to be

989a

that out of which things first come into being by combination, and it is the most finely divided and lightest of bodies that would be of this sort. (For this reason, all those who set up fire as a source would be speaking most in conformity to this argument, but each of the others also agrees that the element of bodies is of this sort. At any rate, none of those who said there was one element thought it fitting for it to be earth, obviously because of its coarse texture, while each of the three elements has won over some judge, for some say this to be fire, some water, and some air. So why in the world don't they also name earth, as

989a 10

most people do? For these do say everything is earth, and Hesiod says that the earth was the first of bodies to come into being—so ancient and popular has this judgment happened to be.) So according to this argument, one could not speak rightly if he were to name any of these except fire, nor if he set up something denser than air but less dense than water. On the other hand, if what is later in its coming into being is more primary in its nature, and what is ripened and compounded is later in its coming into being, the contrary of these things would be the case, and water would be more primary than air, and earth more primary than water.

Now about those people who set up one cause of the sort we are

989a 20

speaking of, let these things we have said be sufficient; but the same thing may also be said if someone sets these up as more than one, as Empedocles says the material consists of four bodies. For in some ways the same things must turn out for him too, as well as others peculiar to him. For we see them come into being out of one another, as though the same body does not always remain fire or earth (but these things were spoken of in the writings about nature), and as for the cause of moving things, whether one ought to set it down as one or two, one ought not to suppose that he spoke either correctly or even completely reasonably. And it is necessary that those who speak this way abolish qualitative change altogether, for cold will not come from hot nor hot from cold. For then something would undergo the opposite conditions themselves, and there would be some one nature which became fire and water, which he denies.

989a 30

And if one assumes that Anaxagoras spoke of two elements, one's assumption would be very much in accord with reason, though he himself did not articulate this; nevertheless it follows necessarily for those who track him down. Now while it is absurd to say that at their source all things are mixed together, not only for other reasons, but also because they would have to turn out to have been present all along as unmixed, and because it is not in accord with nature for any chance thing to be mixed with any other, and on top of these things, because properties and incidental attributes could be separated from the things they belong to (for of the same things of which there is mixing, there is also separating); nevertheless, it one were to follow it up, articulating thoroughly what he wanted to say, he would perhaps appear to be speaking in something more like the latest fashion. For when nothing was separated, it is clear that there was nothing true to say about the thinghood of that mixture; I mean, for example, that it was neither white nor black nor gray nor any other color, but necessarily colorless, since it would otherwise have had a certain one of these colors. Likewise, by this same argument, it was without flavor or any other similar attribute, for it would not be possible for it to be of any sort or of any amount. For then some one in particular of the forms spoken of would have belonged to it, but this is impossible since all are mixed together; for it would already have been separated out, while he says that all things are mixed together except the intellect, which is the only thing unmixed and pure. So from these things it turns out that he is saying that the sources are the one (for this is simple and unmixed), and another which is just like what we set down as the indeterminate prior to its being made definite and participating in some form; so while he speaks neither correctly nor wisely, nevertheless he means something very similar to those who spoke later and who are more manifest now.

989b

989b 10

989b 20

But it is the lot of these people to be at home only with talk about becoming, perishing, and motion, for this is just about the only sort of thing for which they seek the sources and causes. On the other hand, all those who pay attention to all beings, of which some are perceptible and others are not accessible to the senses, obviously make an investigation about both kinds. For this reason one might well spend more time on them, dwelling on what they said well or badly that goes to the inquiry that now lies in front of us.

989b 30   Now those who are called Pythagoreans make use of more remote sources and elements than do those who write about nature. (The reason is that they do not take them from perceptible things, since the mathematical things, apart from the ones involved in astronomy, are without motion.) Nonetheless, they do discourse about and concern

990a   themselves with nature in all their work. For they generate the heavens, and observe what becomes of its parts and attributes and doings, and they lavish their sources and causes on these, as though they agreed with the other writers on nature that what *is* is all this that is perceptible by the senses and surrounded by what is called heaven. But the causes and sources they speak of are, as we say, sufficient to step up also to the higher kinds of beings, and better fitted to this than to their accounts about nature. By what means, however, there will be motion when the limit and the unlimited, or the odd and the even are the only

990a 10   things presupposed, they do not say at all, nor how without motion and change there can be becoming and perishing, or the things done by the things that are carried along through the heavens. What's more, even if one were to grant them that magnitude is made out of these things, or this were demonstrated, still, by what means will there be some bodies that have lightness and some that have heaviness? From the things they assume and say, they are speaking about mathematical things no more than about perceptible ones; for which reason they have not said anything at all about fire or earth or the other bodies of this sort, since, I suppose, nothing that they say about perceptible things is about them in particular.

Again, how ought one to understand number and the attributes of

990a 20   number to be causes of what is and what happens in the heavens, both in the beginning and now, *and* that there is no other kind of number besides this very number out of which the cosmos is composed? For when, say, in some part of the cosmos there is, according to them, opinion and due measure, and a little above or below there is injustice and either separating or mixing, and they claim as a demonstration of this that each of these is a number, but it turns out that in the same place there are already several magnitudes combined since these attributes go along with each of the various places, ought one to understand that this is the same number, this one that is in the heavens, that every one of these is, or another one beside this? For Plato says that it is a different one, even though he too assumes both these things and their causes to be numbers, but the causes to be intelligible numbers, while these other things are sensible numbers.[18]

990a 30

Chapter 9 As far as the Pythagoreans are concerned, then, let the things now said be left alone (for it is enough to touch on them this much). But as for those who set up the forms, first of all, in seeking to understand the causes of the things around us, they brought in other things, equal to these in number, as though someone who wanted to count a smaller number of things thought he couldn't do it, but could count them if he made them more. For the forms are just about equal to, or not fewer than, those things in search of the causes of which they went on from the latter to the former. For over against each particular thing there is something with the same name, and also for the other things besides the beings, to which there belongs a one-over-many, both among the things around us and among the everlasting things.[19] What's more, of those ways by which we show that there are forms, it is not evident by any of them.[20] For from some of them no necessary conclusion results, while from others there turn out to be forms of

990b

990b 10

---

[18] See the note to 987b 23. Sensible numbers are just the multitudes of visible, tangible things. The intelligible numbers meant here are not the mathematical ones, of which the units are all alike, but the eidetic ones, the forms themselves as assemblages of other, unlike forms. The point of this paragraph is that the Pythagoreans had not clarified or even distinguished the inconsistent ways they used the word number.

[19] The manuscripts are hard to make sense of for this sentence. Its point is this: while the forms answering to the species of things around us are fewer than the things are, the number of forms multiplies when they are posited for *every* kind of class characteristic (one-over-many), such as red things, scalene triangles, etc. Aristotle agrees that there are forms in the first sense, but not in the second.

[20] There is a widespread myth according to which Aristotle's teachings are opposite to Plato's. As a result, passages such as this in which he speaks of the disciples of Plato as "we" have to be dismissed as relics of Aristotle's student days, included in his mature writings by some inept editor. This leads to a kind of interpretation as butchery, that hacks apart the text to preserve a point of view that had no sound evidence to begin with. (See the introduction.) Aristotle is among those who posit forms as causes and as beings, and for that reason he is at pains to criticize in detail the technical working out of that doctrine.

those things which we believe do not have them. For as a result of the arguments from the kinds of knowledge, there will be forms of all those things of which there is knowledge, and as a result of the one-over-many there will even be forms of negations, while as a result of the thinking of something that has been destroyed, there will be forms of things that have passed away, since there is an image of these things. On top of this, some of the most precise of the arguments produce forms of relations, which we claim is not a kind in its own right, while other ones imply the third man.[21]

And in general, the arguments about the forms abolish things which we want there to be, more than we want there to be forms. For it turns

990b 20

out that not the dyad, but number, is primary, and that what is relative is more primary than what *is* in its own right, as well as all those things which certain people who took the opinions about the forms to their logical conclusions showed to be opposed to the original sources of things.[22] What's more, by the assumption by which we say there are forms, there will be forms not only of beings but also of many other things. (For the object of the intellect is one thing not only as concerns beings but also as applied to other things, and there is demonstrative knowledge not only about thinghood but also about other sorts of being,[23] and vast numbers of other such conclusions follow.) But both necessarily and as a result of the opinions about them, if the forms are

990b 30

shared in, there must be forms only of independent things. For things do not share in them incidentally, but must share in each by virtue of that by which it is not attributed to an underlying subject. (I mean, for example, if something partakes of the double, then this also partakes of the everlasting, but incidentally, since it is incidental to the double to be everlasting.) Therefore, the forms will be the thinghood of things,

---

[21] The "third man" is shorthand for an argument from an infinite regress which may be found in Plato's *Parmenides*, 132 A–B. All five of the arguments here mentioned are in that dialogue, and all are logical in character; Aristotle's own approach to form is always from nature, or change, or the way the world is.

[22] Platonic doctrine made the one and the dyad sources, but the logical extension of simpleminded one-over-many arguments would force one to derive those sources themselves from the broader class of number that includes them, or from the two-to-one ratio that links them.

[23] This refers to the so-called "categories," the various senses in which something can be said to be, not only as a thing but as a quality, quantity, relation, etc. See 1017a 22–31.

and the same things will signify thinghood there as here. Otherwise,   991a
what would the something be that is said to be apart from the things
around us, the one-over-many? And if the forms and the things that
participate in them have the same form, there would be something
common. (For why is *two* one and the same thing as applied both to
destructible pairs and to those that are many but everlasting, more so
than as applied both to itself and to something?) But if the form is not
the same, there would be ambiguity, and it would be just as if someone
were to call both Callias and a block of wood "human being" while
observing nothing at all common to them.

But most of all, one might be completely at a loss about what in
the world the forms contribute to the perceptible things, either to the   991a 10
everlasting ones or to the ones that come into being and perish. For
they are not responsible for any motion or change that belongs to them.
But they don't assist in any way toward the knowledge of the other
things either (for they are not the thinghood of them, since in that case
they would be in them), nor toward their being, inasmuch as they are
not present in the things that partake of them. In that manner they
might perhaps seem to be causes, after the fashion of the white that is
mingled in white things, but this argument, which Anaxagoras first,
Eudoxus later, and some others made, is truly a pushover (for it is easy
to collect many impossibilities related to this opinion).[24] But surely it
is not true either that the other things are made out of the forms in any   991a 20
of the usual ways that is meant. And to say that they are patterns and
the other things participate in them is to speak without content and
in poetic metaphors. For what is the thing that is at work, looking off
toward the forms?[25] And it is possible for anything whatever to be or
become like something without being an image of it, so that whether
Socrates is or is not, one might become like Socrates, and it is obvious
that it would be the same even if Socrates were everlasting. And there

---

[24] In the *Phaedo,* at 105C, Socrates calls this opinion a "safe but stupid answer" about
the cause of anything. The problem is not that it puts the form into the thing, but
that it offers no adequate conception of what the form is. Aristotle will work that out
in Books VIII and IX.

[25] This is the question out of which Aristotle derives his own account of the forms.
Socrates says in the *Republic* (477d) that we know a potency by its work, using the
same words as are in this sentence, and in the *Sophist,* the Eleatic Stranger defines
being as potency (247E), making the forms causes by ascribing life, motion, and soul
to them, and making them a combination of motion and rest (248E–249D).

would be more than one pattern for the same thing, and so too with the forms; for example, of a human being, there would be animal and at the same time two footed, as well as human being itself. What's more, the forms will be patterns not only of the perceptible things but also of themselves, such as the form *genus,* since it is a genus of forms, and so the same thing would be a pattern and an image.

991a 30

991b

Further, one would think it was impossible for the thinghood and that of which it is the thinghood to be separate: so how could the forms be the thinghood of things if they are separate? In the *Phaedo* it is put this way: that the forms are responsible for both being and becoming. Yet even if there are forms, still the things that partake of them do not come into being if there is not something that causes motion, and many other things do come into being, such as a house or a ring, of which we say there are no forms.[26] So it is clear that the other things too admit of being and becoming by means of the sort of causes which produce the ones just mentioned.

991b 10

Again, if the forms are numbers, how would they be causes? Is it because the beings are various numbers, this number human being, that one Socrates, this other one Callias? Why then are those the causes of these? And it will make no difference if the ones are everlasting and the others not. But if it is because the things here are ratios of numbers, as harmony is, it is clear that there is some one thing of which they are ratios. So if this is something, as material, it is apparent that the numbers themselves will also be particular ratios of one thing to another. I mean, for example, if Callias is a ratio among numbers consisting of fire, earth, water, and air, then the form too will be a number consisting of certain other underlying things, and human

991b 20

being itself, whether or not it is a sort of number, will still be a ratio among numbers *of something,* and not a number, nor would it be on this account a certain number. What's more, from many numbers comes one number, but in what way is one form made of forms? But if it is not made of them but of the things that are in a number, as in ten thousand,

---

[26] Both in Plato's dialogues and in Aristotle's writings a first way in toward the idea of form is the pattern a craftsman has in mind when he makes something. But that example is like a temporary scaffolding that is kicked away once a sounder structure is built. In the *Physics,* the notion that a bed has a form is undermined the instant the idea of being-at-work is introduced (193b 7). Trees have forms, so in a derivative way human artisans can rework the wood.

how does it stand with the units? For if they are homogeneous, many absurdities will follow, and also if they are not homogeneous, either one the same as another or a whole group the same as a whole group; for in what respect will they differ if they are without attributes?[27] For these things are neither reasonable nor in agreement with a thoughtful viewing of the matter.

And on top of this, it is necessary to construct a different kind of number, with which the art of arithmetic is concerned, and all those things that are said by some to be in-between; in what way *are* they and from what sources? Or by what cause are they between the things here and those things? Again, each of the units in the number two comes from some more primary dyad, yet this is impossible. Again, by what cause is a number one thing when it is taken all together? Again, on top of the things that have been said, if the units are to differ, it would behoove one to speak in the same way as those who say there are four, or two, elements. For each of these people speaks of as an element not what is common, such as body, but fire or earth, whether body is something common or not. But as it is, one speaks as though the one were homogeneous, just like fire or water; and if this is so, the numbers will not be independent things. But it is clear that, if there is something that is the one itself, and this is a source, *one* is meant in more than one way, for otherwise it is impossible.

991b 30

992a

992a 10

Now when we want to lead things back to their sources, we set down length as being composed of the short and the long, a certain kind of small and large, and surface of the wide and the narrow, and solid of the deep and the shallow. In what way, though, will the surface have a line in it, or the solid have a line or a surface? For the wide and the narrow are different kinds of thing from the deep and the shallow; so just as number is not present in them either, because the many and the few differ from these, it is clear that neither will any other of the higher things be present in the lower. But surely neither is the wide a class of the deep, for then a solid would be a certain kind of surface. And from what source will the points come to be present in them? Plato used to fight against this class of things as a geometrical dogma, but he called the source of the line—and he set this down often—indivisible

992a 20

---

27 If identity and distinctness are conferred by number, they cannot already be in the units.

lines. And yet there must necessarily be some limit of these; so from the argument from which the line is deduced, the point too is deduced.

In general, though we are seeking wisdom about what is responsible for the appearances, we have ignored this (for we say nothing about the cause from which the source of change is), but supposing that we are speaking about the thinghood of them, we say that there are other independent things, but as for how those are the thinghood of these we speak in vain. For "participating," as we said before, is no help. But on

992a 30   that which we see to be a cause in the various kinds of knowledge, for the sake of which every intellect and every nature produces things, and on that sort of cause which we say is one of the sources of things, the forms do not even touch, but philosophy has turned into mathematical things for people now, though they claim that it is for the sake of other

992b   things that one ought to study them. But still, one might assume that the underlying being which serves as material is too mathematical, and is something to be attributed to or be a distinction within the being or material, rather than being material; for instance, the great and the small are just the same as what the writers on nature call the rare and the dense, claiming that these are the first distinctions within the underlying material, since they are a certain kind of the more and less. And as for motion, if these are a process of moving,[28] it is clear that the forms would be moved; but if they are not, where does it come from? For then the whole investigation about nature would be abolished.

992b 10   And what seems to be easy, to show that all things are one, does not happen; for from the premises set out, even if one grants them all, it does not come out that all things are one, but that something is the one itself; and not even this comes out if one will not grant that the universal is a genus, which in some cases is impossible.[29] And there is no explanation of the lengths, surfaces, and solids that are present with the numbers, not of how they are or could be nor of what capacity they have; for it is not possible for them to be forms (since they are not

---

[28] That is, if the great and small refer to a process of becoming bigger or smaller.

[29] For example, being is not the genus of all the things that are, but several irreducible genera, of which one is primary. Likewise, good is not the genus of all good things, which are the same only by analogy. The understanding of these fruitful ambiguities is at the heart of the argument of the *Metaphysics,* and begins to be worked out in Bk. IV, Ch. 2.

numbers), nor the in-between things (since those are the mathematical ones), nor perishable things, but this seems in turn to be some other fourth kind.

In general, to search for the elements of whatever *is* without distinguishing the many ways this is meant, is to seek what is impossible to find, both for those inquiring in other ways and those inquiring in this way about what sort of elements things are made of. For out of what elements acting is made, or being acted upon, or the straight, is just not there to be grasped, but if of anything, it is of independent things alone that it can possibly be. Therefore to suppose that one is seeking or has the elements of *all* beings is not true. And how could anyone learn the elements of all things? For it is clear that it is not possible for someone to begin by knowing it beforehand. For just as it is possible for the one who is learning to do geometry to know other things in advance, though he does not already know any of the things of which geometry is the knowledge and about which he is going to learn, so it is also with other things, so if there is any knowledge of all things, of the sort that some people say there is, he could not start out already knowing anything.[30]

992b 20

992b 30

And yet all learning is by means of things all or some of which are already known, whether it is by means of demonstration or by way of definitions (for it is necessary that one already know and be familiar with those things out of which the definition is made), and likewise with learning through examples. But if it happens to be innate, it is surely a wonder how we fail to notice that we have the most powerful kind of knowledge. Furthermore, in what way will one know what something is made of, and how will it be evident? For this too contains an impasse: for one might dispute in the same way as about some syllables. For some people say that *za* is made of *s*, *d*, and *a*, while some others say that it is a distinct sound and not made of any familiar ones. What's more, how could someone know those things of which sense perception consists if he did not have the sense capacity? And yet he must, if indeed the elements out of which all things are made are the same, in just the way that composite sounds are made of the elements that belong to them.

993a

993a 10

---

[30] This argument does not reject the idea of recollection but presupposes it. If wisdom is conceived in a too-unified way, it leaves over nothing out of which we could begin to recollect.

Chapter 10 That, then, everybody seems to be looking for the causes spoken of in our writings on nature, and that outside of these there is none that we could speak of, is evident even from what has been said above. But they inquired murkily into these, and while in a certain way all the causes have been spoken of before, in another way they have not been spoken of at all. For the earliest philosophy about everything is like someone who lisps, since it is young and just starting out. Even Empedocles said that bone *is* by means of a ratio, and this is what it is for it to be and the thinghood of the thing.[31] But surely also flesh and each of the other things would have to be through a ratio in the same way, or not even one thing. Therefore it is through this that flesh and bone and each of the other things would have being, and not, as *he* says, through the material, the fire, earth, water, and air. But while he would necessarily concede these things if someone else said them, he did not say them clearly. Now what concerns these things has also been made clear above; but let us go back again over all those things that anyone might find to be impasses about these same topics, for perhaps out of them we might provision ourselves in some way for the impasses that will come later.

993a 20

---

[31] In fragment 96 (Diels's numbering), Empedocles speaks of bone being made by the goddess Harmony as one-fourth earth, one-fourth water, and one-half fire. So while he had no grasp of form as a cause, his words implied it.

Book II (Book α)[1]
Inquiry[2]

Chapter 1   The beholding of truth is in one way difficult, but in 993a 30
another way easy. A sign of this is that, while no one happens to be
capable of it in an adequate way, neither does anyone miss it, but each 993b
one says something about nature, and though one by one they add
little or nothing to it, from all of them put together something comes
into being with a certain stature. So if it seems that we happen to be in
the condition of the common saying, "who could miss the doorway?,"
in this way it would be easy, but to have the whole in a certain way,
and yet be incapable of part of it, shows what is difficult about it. But
perhaps, since the difficulty is of two sorts, the cause is not in the
things but in us; for in just the way that the eyes of bats are related
to the light of mid-day, so also is the intellect of our soul related to 993b 10
those things that are by nature the most evident of all. And it is right
to feel gratitude not only to those whose opinions one shares, but
even to those whose pronouncements were more superficial, for they
too contributed something, since before us they exercised an energetic
habit of thinking. For if there had been no Timotheus, there is much
lyric poetry we would not have had, but were it not for Phrynis, there
would have been no Timotheus. And it is the same way too with
pronouncements about truth; for we have inherited certain opinions
from certain people, but others have been responsible for bringing
them about.

   And it is also right to call philosophy the knowledge of truth. For the 993b 20
end of contemplative knowledge is truth, but of practical knowledge
it is action; for even if people devoted to the active life do examine the
way things are, they do not contemplate the cause in its own right, but

---

[1] This book is labeled "little alpha," and is an obvious insertion between Books I and III.
The *Metaphysics* is pieced together from many separate strands of related writings, but
it is assembled with great care to form a single argument. There are some overlapping
passages, and others not strictly necessary to the dialectical advance of the whole, but
no open-minded reading could confirm the judgement of those scholars who think
the parts represent different and incompatible "stages" of Aristotle's "development."
The coherence of the work is not apparent by philological analysis, but emerges
unmistakably for a serious philosophic reader. (See the Introduction.)

[2] This title for Book II supplied by the translator.

in relation to something now. But without the cause we do not know the truth, but each thing is what it is most of all, and more so than other things, if as a result of it the same name also belongs to those other things. (For example, fire is the hottest thing, since it is also responsible for the hotness of the other things.) Therefore also, what is responsible for the being-true of derivative things is more true than they are. For this reason the sources of the things that always *are* must be true in the highest sense (for they are not sometimes true, nor is anything a cause

993b 30   for them of their being, but they are the cause of other things), so what each thing has of being, that too it has of truth.[3]

994a   **Chapter 2**   Now surely it is clear that there is some source of things and that the causes are not infinite either in a straight line or in kind. For it is not possible for one thing to come from another infinitely, either as from material (such as flesh from earth, earth from air, air from fire, and in this way without stopping), or from where the origin of motion is (such as for a human being to be moved by the air, this by the sun, the sun by strife, and for there to be no limit of this); nor, similarly, can that for the sake of which go on to infinity—walking for

994a 10   the sake of health, this for the sake of happiness, happiness for the sake of something else, and for one thing to be for the sake of another forever in this way—and likewise in the case of what it is for something to be. For of in-between things, which have some last thing and some more primary one, the more primary must be the cause of the ones after it. For if we had to say which of the three things was the cause, we would say the first one; for it is surely not the last one, since the last of all is the cause of nothing, but neither is it the middle one, though it is the cause of one thing (and it makes no difference if there is one or more than one middle thing, nor whether there are infinitely or finitely many). But of things infinite in this way, and of the infinite in general, all the parts are alike middle ones down to the present one; therefore, if there is no first thing, there is no cause at all.

994a 20   But surely it is not possible to go to infinity in the downward direction either, of something that has a beginning above, so that out

---

[3] Etymologically, the Greek word for truth means something like "what emerges from hiddenness." It therefore applies to things as well as to knowing. This paragraph is a first and highly compressed sketch of the structure of the *Metaphysics* as a whole. It is a search for that which most of all *is,* as a result of which other things are.

of fire would come water, and out of this earth, and so forever some other kind coming into being. For one thing comes out of another in two ways: either as a man comes into being out of a boy by his changing, or as air comes into being out of water. So the sense in which we say a man comes into being out of a boy is as what has become comes out of what is becoming, or what is complete out of what is being completed. (For just as becoming is always between being and not being, so also the thing that is becoming is between something that is and something that is not; for the one who is learning is a knower coming into being, and this is what it means to say that a knower comes out of a learner.)   994a 30
But the sense in which air comes into being out of water is by the destruction of one of the two things. For this reason the former kind do not turn back into one another, and a boy does not come into being out of a man (for out of the process of becoming there does not come   994b
something that is becoming, but something that is after the process of becoming, for so too a day comes out of morning because it is after it, for which reason morning does not come out of a day), but the other kind do turn back into one another. But in both ways it is impossible to go to infinity; for of the in-between sort of beings there must necessarily be an end, while the other sort turn back into one another, since the destruction of one of them is the coming into being of the other. But at the same time it is impossible for the first thing, which is everlasting, to be destroyed; for since the process of becoming is not infinite in the upward direction, what came into being out of a first thing that was destroyed could not be everlasting.[4]

And since that for the sake of which something *is* is an end, and this sort of thing is what is not for the sake of anything else, but they for the   994b 10
sake of it, then if there is any such last thing, there will not be an infinity, but if there is no such thing, there will be nothing for the sake of which it is. But those who make there be an infinite are unaware that they abolish the nature of the good. (Yet no one would make an effort to do anything if he were not going to come to a limit.) And there would not be intelligence among beings; for what has intelligence always acts for the sake of something, and this is a limit. And neither is it possible that

---

[4] The last sentence means that even if a process of becoming began with, say, water, and water never reappeared in the series, still the very fact that it was destroyed would mean it could not have generated an infinite series of effects. Compare *Physics* 266a 10–266b 6.

what it is for something to be should be led back to another definition
that has a fuller articulation; for it is always the earlier definition that

994b 20   is more and the later that is not, and what is not in the first one is not
in the next one either.[5] What's more, those who say that it is abolish
knowing, since it is not possible to know until one has come to what is
indivisible; and there isn't any knowing, for how is it possible to think
what is infinite in that way? For it is not the same thing as with a line,
which does not come to an end of divisions but which is not possible
to think unless one stops dividing it (for which reason the one who
goes through the line as an infinity will not count up the divisions).

But in something that moves it is also necessary to think about the
material. And for nothing is it possible that it be infinite; or if it is, at
least the being-infinite of it is not an infinite series.[6] But surely also if
the kinds of causes were infinite in number, neither would there be any
knowing for that reason; for we think that we know something when

994b 30   we are acquainted with its causes, but what is infinite by addition is
not possible to go all the way through in a finite time.

Chapter 3   Courses of lectures go along with one's habits; for in
995a   the way that we are accustomed, in that way we think it fitting for
something to be said, and what departs from this does not seem the
same, but through lack of acquaintance seems too obscure and alien.
For what we are used to is familiar. And what great strength the
customary has, the laws show, in which the mythical and childish
things are of greater strength than knowing about them, because of
custom. Some people do not give a favorable reception to what is said
if one does not speak mathematically, others if one speaks without
giving examples, and others expect one to bring in a poet as a witness.
Some expect everything to be said with precision, while others are
995a 10   annoyed by precision, either because they can't keep the connections

---

[5] By "what it is for something to be," Aristotle means (a) what a thing always is, or
continues being, and (b) all that is present in it in order *that* it be. The first qualification
excludes what is individual and incidental, while the second excludes what is universal
and incomplete. The species is at the exact rung of the scale of particularity and
generality that articulates what anything is, and one does not improve it by adding
something from either side. Only self-sustaining independent beings have this kind of
articulation.

[6] Even if the material of the world were infinite in extension it would not constitute
an infinity of material causes.

straight or because of its hairsplitting pettiness. For precision does have something of this sort about it, so that, just as in business agreements, so also in reasoning it seems to some people to be ungenerous. For this reason one must have been trained in how one ought to receive each kind of argument, since it is absurd to be searching at the same time for knowledge and for the direction to knowledge; and it is not possible to get either of the two easily. Now mathematical precision of speech is something one ought to demand not in all things, but concerning those that do not have material. For this reason it is not a way that is suited to nature, for presumably all nature has material. Therefore one must first consider what nature is, for in this way it will also be clear what the study of nature is concerned with.[7]

---

[7] This requirement is not picked up immediately, but is part of an array of preliminary cautions in the woven texture of the *Metaphysics*. It is picked up indirectly at two of the main structural points in the work: at 1029a 33–1029b 12, where the inquiry gets going in earnest, and at 1064a 10–28, where a passage from the *Physics* is inserted to prepare the way to the culmination of the inquiry in Book XII.

Book III (Book B)
Impasses[1]

Chapter 1 It is necessary, looking toward the knowledge that is being sought after, for us first to go over those things about which one must first be at an impasse. And these are all those things about these topics that some people have conceived in different ways, as well as anything apart from these that they might happen to have overlooked. And it is profitable for those who want to get through something well to do a good job of going over the impasses. For the later ease of passage is an undoing of the things one was earlier at an impasse about, but it is not possible to untie a knot one is ignorant of. But the impasse in our thinking reveals this about the thing, for by means of that by which one is at an impasse, one suffers in much the same way as people who are tied up, for in both cases it is impossible to go on forward. For this reason it is necessary to have looked at all the difficulties beforehand, both on these accounts and because those who inquire without first coming to an impasse are like people who are ignorant of which way they need to walk, and on top of these things, because one never knows whether one has found the thing sought or not. For the end is not apparent to this one, but to one who has first been at an impasse it is clear.[2] And further, one must be better off for judging if one has heard all the disputing arguments as if they were opponents in a lawsuit.

The first impasse is about the things that we raised questions about in the prefatory chapters, whether it belongs to one or to many kinds of knowledge to contemplate the causes, and whether it is only the first sources of being at which it belongs to this knowledge to look, or whether it is also concerned with the starting points from which everyone demonstrates, such as whether it is possible or not to assert and deny one and the same thing at the same time, and with other things of this sort; and if it is about being, whether one knowledge or more than one concerns all beings, and if more than one, whether all

---

[1] This title for Book III supplied by the translator.

[2] This passage should be compared with the crisis in Plato's *Meno,* at 79E–81A, which is an impasse about impasses.

are akin or some are kinds of wisdom while others of them must be called something else. And this itself is one of the most necessary things to inquire about: whether one ought to say that there are perceptible things only or also others besides these, and whether the latter are of one kind or there is more than one kind of independent things, as those say who make there be both the forms and the mathematical things, between these and the perceptible things. So as we say, one must investigate about these things, and also whether the examination

995b 20   is about beings only or also the attributes that belong to beings in virtue of themselves; and on top of this, about same and other, like and unlike, contrariety, prior and posterior, and all the other things of this sort about which those engaged in dialectic try to make an inquiry by considering only what follows from accepted opinions, one must investigate what endeavor is concerned with examining all these, and what's more, all that is attributed to these things themselves in their own right, and not only what each one of them is, but also whether one is contrary to one. And one must consider whether the sources and elements are the kinds of things or the constituents into which each thing is divided, and if they are the kinds, whether they are the last

995b 30   ones or the first ones attributed to the individual things, for example whether animal or human being is a source and is more separate from the particular thing.

But most of all, one must inquire and exert oneself about whether there is or is not anything apart from material that is a cause in its own right, and whether this is something separate or not, and whether it is one or more than one in number, and whether there is anything apart from the composite whole (and I speak of a composite whole whenever something is predicated of the material), or nothing apart from it, or whether there is something apart from some of them but not from others, and what sort of beings would be of this

996a   kind. Further, are the sources, both of the articulation of things and of what underlies them, limited in either number or kind? And of destructible and indestructible things are they the same or different, and are they all indestructible or are those of destructible things destructible? Furthermore, the most difficult question of all, that has in it the greatest impasse, is whether *one* and *being,* as the Pythagoreans and Plato said, are not anything different, but are the thinghood of things—or whether this is not so, but the underlying thing is something different, friendship, as Empedocles says, or someone else fire but

another water or air. And there is a question whether the sources of
things are universal or like particular things, and whether they have
being potentially or at work, and in turn whether they are at work in
some other way or by way of motion; for these would also present a
considerable impasse. On top of these things, are numbers and lengths
and shapes and points certain independent things or not, and if they
are independent things, are they separated from perceptible things or
constituents in them? About all these things it is not only difficult to
find a way to the truth, but it is not even an easy thing to articulate the
difficulties well.

Chapter 2 First, then, are the things about which we spoke first,
whether it belongs to one or to more kinds of knowledge to contemplate all the kinds of causes. For how could it belong to one kind of
knowledge to know sources that are not contraries? What's more, to
many beings, not all the causes belong; for in what way is a source
of motion possible in motionless things, or the nature of the good, if
everything that is good in itself and through its own nature is an end,
and a cause in the sense that for its sake other things come into being
and are, and the end and that for the sake of which is an end for some
action, while all actions include motion? Therefore among motionless
things there could not be this kind of source, nor could there be any
good-itself. For this reason too, among mathematical things nothing is
demonstrated by means of this kind of cause, nor is there any demonstration on account of better or worse, but no one pays any attention at
all to anything of that sort; and so on this account some of the sophists,
such as Aristippus, used to belittle mathematical demonstrations. For
in the other arts, and in the mechanical skills such as carpentry and
shoemaking, everything is accounted for because of the better and
worse, while mathematics makes no mention about what is good and
bad.

But now if there are a number of kinds of knowledge of the causes,
and a different one for a different source, which of these ought one to
say is the one being sought, or who among those who have them is
most of all a knower of the thing that is being sought? For it is possible
for all the kinds of causes to belong to the same thing; for example, with
a house, that from which the motion comes is the art or the builder,
that for the sake of which is its work, the material is earth and stones,
and the form is its articulation in speech. From the distinctions made

996b 10   just now about what kind of knowledge is wisdom, there is reason to apply the name to each one. For in the sense that it is the most ruling and leading and that, like slaves, it is not even fitting for the other kinds of knowledge to talk back to it, the knowledge of the end or the good is of this sort (since the other things are for the sake of this), but in the sense that it was defined as being about the first causes and the most knowable thing, the knowledge of thinghood would be of that sort. For of those who know the same thing in various ways, we say that he knows it more who is acquainted with the *what* of the thing by what it is, rather than by what it is not, and among those with the former knowledge themselves, one knows it more than another, and most of all, the one who knows what it is, rather than its size or quality or how it acts or is acted upon by nature. And also among other things,

996b 20   about which there are demonstrations, we think that the knowing of each of them is present at the time when we see what it is (for example, what squaring is, that it is finding a mean proportional,[3] and similarly in other cases), but about instances of coming into being and actions and about every change, at the time when we see the source of the motion, and this is different from and opposite to the end; therefore, the contemplation of each of these causes would seem to belong to a different kind of knowledge.

But now also about the demonstrative principles, it is a matter of dispute whether they belong to one or more than one kind of knowledge. By demonstrative principles I mean the common opinions from which every one demonstrates, such as, that everything must be

996b 30   either asserted or denied, and that it is impossible for something both to be and not be at the same time, and as many other such premises as there are. Is there one kind of knowledge about these as well as about thinghood,[4] or different ones, and if there is more than one, to which

---

[3] We think of "squaring" as a procedure of arithmetic, but its original meaning is the construction of a square equal to a given oblong rectangle. The side of the square will have the same ratio to one side of that rectangle that the other side has to it. One can know how to find the side of the square without understanding that, as in Euclid's proposition II, 14; in Euclid's *Elements* the understanding of that connection comes only in proposition VI, 13. This example of two ways of knowing is used by Aristotle at the beginning of Bk. II, Ch. 2, of *On the Soul*.

[4] In this passage, the word *ousia* is used in three ways: it may refer to beings as opposed to principles or propositions, or it may single out independent things as opposed to the various kinds of attributes that are all beings in the widest sense, and it also means

one ought one to give the name of the thing now being sought? Now it is not reasonable that they belong to one; for why is it peculiar to this kind of knowledge more than to geometry or any other whatever to pay attention to these principles? So if it pertains in the same way to any whatever, but cannot belong to them all, then just as it is not peculiar to the others, so also it is not peculiar to the one that explains independent things to know about them. But by the same token, in what way could there be a kind of knowledge about them? For even now we are familiar with what each of them is. (At any rate even the other arts use them as though familiar with them.) But if there is a demonstrative knowledge about them, there would need to be some underlying kind, while some of them are attributes and others of them are axioms (since it is impossible for there to be a demonstration about everything), for the demonstration must be from some things, about some topic, and of some things; therefore it follows that there would be some one class of all things that are demonstrated, since all demonstrative knowledge makes use of the axioms. But now if the knowledge of thinghood and that about these principles are different, which of them is by nature more authoritative and more primary? For the axioms most of all are universal, and are starting points of all things, and if it does not belong to the philosopher, to whom else would it belong to consider what is true and false about them?

997a

997a 10

And in general, is there one or more than one kind of knowledge about all beings? And then if there is not one, with what sort of beings ought one to place this kind of knowledge? But that there is one about them all is not reasonable; for then there would also be one kind of demonstrative knowledge about all attributes, if every kind of demonstrative knowledge about some underlying subject examines the attributes it has in its own right on the basis of common opinions. So it concerns the same class of things to examine on the basis of the same opinions the attributes it has in its own right. For that about which it demonstrates belongs to one kind of knowledge, and those things from which it demonstrates belong to one kind of knowledge,

997a 20

---

thinghood, the way of being an independent thing, or the cause responsible for that condition. The common translation "substance" not only misses all these meanings, but carries implications of its own that can only mislead and confuse. For more on this, see the Introduction and Glossary.

whether the same one or another, so that either these, or one composed of these, will examine the attributes.

But still, is the examination about independent things alone, or also about the attributes of these? I mean, for example, if a solid is an independent thing, and also lines and planes, whether it belongs to the same kind of knowledge to know these and the attributes that pertain to each class, about which mathematics demonstrates, or to a different one. For if it belongs to the same one, there would be a certain demonstrative knowledge even about thinghood, but it does not seem there could be a demonstration of what something is; but if it belongs to a different one, what would it be that examines the attributes that pertain to thinghood? For this is exceedingly difficult to give an account of.

997a 30

Again, ought one to say that there are perceptible things only, or also others besides these, and do there turn out to be classes of independent things in only one or in more than one way, as those say who speak of the forms and the in-between things, about which they say the mathematical kinds of knowledge are? Now the way that we say the forms are both causes and things in their own right has been explained in our first arguments about them; but in a number of ways these are hard to swallow, none being less absurd than saying that there are certain natures apart from those within the heavens, and yet to say that these are the same as the perceptible things except that the ones are everlasting and the others destructible. For they say there is a human being itself, and a horse itself, and a health itself, but say nothing else, doing just about the same thing as those who say there are gods, but having human form. For those people made nothing other than everlasting human beings, and these do not make the forms anything but everlasting perceptible things.[5]

997b

997b 10

But further, if one posits, besides the forms and the perceptible things, the in-between things, there will be many impasses; for it is clear that there would be lines apart from lines-themselves and perceptible lines, and similarly with each of the other classes. Therefore, since astronomy is one of these classes, there will also be a certain heaven apart from the perceptible heaven, and a sun and a moon and

---

[5] Note that the absurdity is not positing forms, but failing to think through how they differ from that of which they are the forms. Aristotle is not quarreling with attaching "itself" to human being, but with saying no more than that.

likewise the other things throughout the heavens. Yet how is one supposed to believe these things to be? For it is not reasonable that they be motionless, but that they be in motion is completely impossible. And it is similar with those things about which optics concerns itself, and also the study of harmony within mathematics. For it is also impossible that these be apart from the perceptible things on account of the same causes; for if there are perceptible things and sense perceptions in-between, it is clear that there would also be animals between the animals-themselves and the perishable ones.[6] But one might also be at an impasse about what sort of things these kinds of knowledge are supposed to inquire after. For if geometry differs from land surveying only in this, that one of them is about things we perceive and the other is not about perceptible things, it is clear that there would also be some sort of knowledge besides the art of medicine and besides each of the other arts, between medicine itself and the medicine here; yet how is this possible? For there would also be some healthy things besides the perceptible ones and the healthy itself. But at the same time not even this is true, that land surveying is about perceptible and perishable magnitudes, for then it would be destroyed when they were destroyed. But surely neither could astronomy be about perceptible magnitudes nor about this heaven. For neither are perceptible lines the sort of lines about which the geometer speaks (for none of the perceptible things is straight or round in that way, for a ring touches a straight edge not at a point but in the way that Protagoras used to say when refuting the geometers[7]), nor are the motions and loops of the heavens the sort of things about which astronomy makes its arguments, nor do points have the same nature as the stars.

But there are some people who claim that there are these things that are said to be between the forms and the perceptible things, but that they are not separate from the perceptible things but in them; to go through all the impossible consequences of this would be too much talk, but it is enough to consider even such ones as these. For it is not

997b 20

997b 30

998a

998a 10

---

[6] It was assumed that all mathematical objects are between the forms and the perceptibles, but optics and harmonics are mathematical disciplines whose objects are perceptible things.

[7] One may find the same sort of arguments, with the same purpose, in the writings of a modern Protagorean, David Hume. See *A Treatise of Human Nature,* Bk. I, Pt. II, Sect. IV.

reasonable that it would be thus with these only, but it is clear that the forms too would admit of being in the perceptible things (for both of these have the same argument), yet two solids would have to be in the same place, and they could not be motionless if they were in moving perceptible things. And in general, for what purpose would anyone posit that they *are*, but are in the perceptible things? For the same absurdities would follow upon the aforementioned ones: there would be a heaven apart from the heaven, except not separate but in the same place, which is even more impossible.[8]

998a 20   **Chapter 3** About these things, then, there is a considerable impasse as to how one must set them down in order to come upon the truth, and also about the sources, whether one must assume that the classes of things are elements and sources, or rather those things out of which, as first constituents, each thing is made. For example, the elements and beginnings of speech seem to be those first things out of which articulate utterances are composed, but not the common genus, speech; and of geometrical constructions, we call elements those things of which the demonstrations are already present in the demonstrations of the others, either of all or of most of them. Again, of bodies, both those who say there is more than one element and those who say there

998a 30   is one say that the things of which they are composed are the sources; for example, Empedocles says that fire and water and the elements that go with these are the things out of which, as constituents, beings are made, but not that these are classes of beings. And on top of these

998b   things, if anyone wants to look into the nature of anything else, say of a bed, it is when he knows what parts it is composed of and how they are put together that he knows its nature. So from these arguments the classes of things would not be the sources of them, but if we know each thing by means of definitions, and if the classes are the sources of the definitions, then the classes must also be the sources of the things defined. And if getting hold of the knowledge of things is grasping the species in accordance with which they are spoken about, the genus is certainly a source of the species. And some of the people who speak of

---

[8] See the additional arguments given in the first paragraph of Bk. XIII, Ch. 2, and the footnotes there. The reader may follow those notes to Aristotle's own account of mathematical things; he agrees that they are not separate from perceptible things, but does not grant that they are in them.

the one or being or the great and the small as elements of things seem to   998b 10
use them as classes. But surely it is not possible to speak of the sources
of things in both ways. For there is a single articulation of thinghood,
while the definition by means of classes would be different from that
which states the things out of which, as constituents, something is
made.

But on top of these things, even if the classes are sources as much
as one could wish, should one regard the first classes as sources, or
the last ones that are applied to the individual things? For this too is
a matter of dispute. For if it is always the universal things that are
more the sources, it is clear that these will be the highest of the classes,
since these are predicated of everything. Then there would be as many
sources of things as there are primary classes, so that being and oneness   998b 20
would be sources and the thinghood of things, since these most of all
are predicated of all things. But it is not possible for either oneness
or being to be a single genus of things. For it is necessary for each
of the things that differentiate each genus to be and to be one; but it
is not possible either to predicate the species within a genus of their
own differentiae, or to predicate the genus without its species of the
differentiae.[9] Therefore, if oneness or being is a genus, no differentia
would either be or be one. But surely if they are not genera, they will
not be sources either, if the genera are sources. What's more, the in-
between classes, understood as including their differentiae, would be
genera down to the individual things (but as things are, some do, others   998b 30
do not, seem to be so). And on top of these things, the differentiae are
sources still more than are the genera, but if these too are sources, there
would come to be (one might say) an infinity of sources, whether or
not one sets down the first genus as a source.   999a

But now if what is one is in fact more sourcelike, and what is
indivisible is one, and everything that is indivisible is so either in
amount or in kind, and indivisibility in kind is more primary, and

---

[9] If we define doves as wild pigeons, the species is doves, the genus pigeons, and the
differentia is being wild. If this is a sound definition, it cannot be true that (all) wild
things are doves, or, the more important point here, that (all) wild things are pigeons.
The reason is that all characteristics by which a genus is differentiated into species are
outside the genus. Hence there cannot be any genus that includes all things, and being
cannot be understood as the class of all beings. Parmenides confirms this: if being has
only one all-comprehensive meaning, there can only be one being. Aristotle's reply
begins in Bk. IV, Ch. 2.

general classes are divided into species, then the lowest predicate would be more of a unity; for human being is not a genus of particular human beings. What's more, in those things in which there is a prior and a posterior, what is in these cannot be something apart from them. (For example, if two is the first of the numbers, there will not be some number apart from the kinds of numbers, and likewise no shape apart

999a 10   from the kinds of shapes; and if there are none among these, there would hardly be any general classes apart from the species of other things, for of the former there seem most of all to be genera.) But among individual things, there is not one prior and another posterior. Yet wherever one thing is better and another worse, the better is always more primary, so that there would not be a genus of these. So from these things, it seems that the predicates applied directly to the individual things are sources more than are the general classes; but then in turn, in what way one ought to understand these to be sources is not easy to say. For the source and cause has to be distinct from the things of which it is the source, and has to be capable of being when separated

999a 20   from them; but why would anyone assume that there was any such thing apart from the particulars, except because they are predicated as universals and of all things? But then if it is for this reason, the things that are more universal must be set down more so as sources; therefore the first classes would be sources.

Chapter 4   But following upon these things is an impasse that is both the most difficult of all and the most necessary to examine, close by which the argument has now come to stand. For if there is nothing apart from the particular things, while the particulars are infinite, how is it possible to get hold of a knowledge of infinitely many things? For insofar as something is one and the same, and insofar as something is present as a universal, in this way we know everything. But if this is necessary, and there has to be something apart from the particulars,

999a 30   the general classes of things would have to have being apart from the particulars, either the lowest or the highest classes. But we just went through an argument that this is impossible. Still, if as much as one wants there is something apart from the composite (and I speak of a composite whenever anything is predicated of the material), must

999b   there be something answering to all of them, or to some of them and not to others, or to none?

Now if there is nothing apart from the particulars, there could be nothing intelligible, but everything would be perceptible and of nothing could there be knowledge, unless someone claims that sense perception is knowledge. What's more, neither could there be anything everlasting or motionless (since all perceptible things pass away and are in motion). But surely if there is nothing everlasting, neither could there be coming-into-being. For there must be something that comes into being and something out of which it comes into being, and the last of these must be ungenerated, if the series comes to a stop and it is impossible to come into being from what is not; but if there is becoming and motion, there must also be a limit (for there is no infinite motion,   999b 10 but of every one there is a completion, and something cannot be coming into being if it is impossible for it to have come into being; but what has come into being must necessarily *be* when it has first come into being). Further, if the material has being because it is ungenerated, much more still is it reasonable that the thinghood has being, which at any time the former is becoming. For if there should be neither thinghood nor material, nothing would be at all, but if this is impossible, there must necessarily be something apart from the composite whole, namely the form or species. But if one sets this down as one's opinion, there is in turn an impasse as to what cases one sets it down in and what cases not. For that it is not so in all cases is clear; for we would not decide that there was any house apart from the particular houses. And besides   999b 20 this, will the thinghood of all the things, say all human beings, be one? But that's absurd, for all things of which the thinghood is one are one being. But is the thinghood many and different? This too is illogical. And at the same time, how does the material become each of these things, and how is the composite both of the two?

Further, one might also be at an impasse about the sources in this way: for if it is in *kind* that they are one, nothing will be one in number, not even one-itself or being-itself. And how will there be knowing, if there is not something that is one in all the particulars? But surely if it is one in number, and each of the sources is one—and not, as with perceptible things, different ones for different things (for example, since such-and-such a syllable is the same in kind, the sources of it   999b 30 are also the same in kind, for these too are present as numerically different)—if, then, the sources of things are one not in that way but in number, there would not be anything else besides the elements. For what is one in number means nothing different from what is particular; for we speak of the particular in that way, as one in number,

1000a   but the universal as what applies to these. So just as, if the letters of
language were definite in number, all writing would necessarily be
just so much as the letters, there could not be two or more of the
same beings.

And an impasse no lesser than any has been neglected by both
present and earlier thinkers, as to whether the sources of destructible
and indestructible things are the same or different. For if they are
the same, in what way and through what cause are some things
destructible and others indestructible? Now some people associated
1000a 10   with Hesiod and all those who gave accounts of the gods considered
only what was persuasive to themselves, but took no heed of us. (For
making the sources be gods and things begotten by gods, they say that
what has not tasted of nectar and ambrosia becomes the mortals; it is
clear that these names they speak of are familiar to themselves, and
yet even about the very introduction of these causes they have spoken
past us. For if the gods have contact with them for the sake of pleasure,
the nectar and ambrosia are not at all responsible for their being, but if
they are responsible for their being, how could they be everlasting and
yet need food?) But about mythological subtleties it is not worthwhile
1000a 20   to inquire seriously; but on the part of those who speak by means of
demonstrations, one must learn by persistent questioning why in the
world, when things come from the same sources, some of the things
have an everlasting nature but others pass away. But since they neither
state any cause, nor is it reasonable that it be so, it is clear that there
could not be the same sources or causes of them.

For even he whom one might suppose to have spoken most consistently with himself, Empedocles, has also suffered the same thing.
For he sets down strife as a source responsible for destruction, but
nonetheless it would seem that this also generates things, except for
the *one*; for all the other things except the god come from this. At any
rate, he says

Out of them came everything, all that was and all that is;
1000a 30   Trees sprouted, and men and women too, and
Beasts and birds and water-nurtured fish, and
Gods living long ages.
1000b   And it is clear apart from these things; for if strife were not present
in things, everything would be one, as he says. For whenever things
come together, then "strife stands farthest away." And for this reason
it follows for him that the most blessed god is less wise than others

since he does not know everything, for he does not have strife, but knowing, Empedocles says, is of like by like.

> For by earth we behold earth, by water, water, and by ether, godlike
>   ether,
> By love, love, and strife by miserable strife.

But that from which this account started is clear: that it turns out for him that strife is responsible for destruction no more than for being. And similarly, neither is friendship more responsible for being, for by bringing things together into one it destroys other things. And at the same time, about a cause of change itself, he says nothing except that that's the way it is by nature.

1000b 10

> But when strife had grown great in its limbs, and
> Leapt up into honor when the time was complete,
> Which is forced on them in turns by a broad-shouldered oath...

as though it were a necessity to change; but he shows no cause at all of the necessity. Nevertheless, only he speaks consistently at least to this extent: he does not make some beings destructible and others indestructible, but all of them destructible except the elements. But the impasse now being discussed is why some things are destructible and other are not, if they come from the same sources.

1000b 20

Then to the effect that there could not be the same sources, let so much have been said; but if there are different sources, one impasse is whether they themselves would be indestructible or destructible. For if they are destructible, it is clear that they too must be derived from something (since everything that passes away passes into those things out of which it is made), so that it follows that there would be other, more primary, sources of the sources; but this is impossible, whether the series stops or goes on to infinity. What's more, how will there be destructible things, if their sources are done away with? But if they are indestructible, why will destructible things come from some of these things that are indestructible, but indestructible ones from others of them? For this is not reasonable, but is either impossible or needs a lot of explanation. Furthermore, no one has even tried to speak about different sources, but all say that the same sources belong to all things. But they gulp down the thing first stated as an impasse as though taking it to be something small.

1000b 30

1001a

But the most difficult thing of all to examine, as well as the most necessary for knowing the truth, is whether being and oneness are the thinghood of things, each of them not being anything else but oneness

or being, or whether one must inquire what it is that is or is one on
the assumption that they belong to some other underlying nature. For
some people suppose that they have the former sort of nature, others

1001a 10   the latter. For Plato and the Pythagoreans suppose that neither being
nor the one is any other thing, but that this is the nature of them, that
the thinghood of each is its being-one or being-being. The others are
the writers about nature; for example, Empedocles, as though tracing
it back to something more knowable, says what the one is. For he
would seem to be saying some such thing as that it is friendship (at
any rate this is the cause of the being-one of everything), but others
say of fire, still others of air, that this is the unity and the being out
of which things are made and from which they have come into being.
And it is the same way also with those who set down the elements as
more than one, for they have to say that oneness and being consist of

1001a 20   just so many things as they say are sources. But it follows, if one does
not posit that oneness or being is an independent thing, that none of
the other universal things is an independent thing either. (For these
are the most universal of all things, but if there is not any one-itself
or being-itself, there could hardly be any of the other universals apart
from the particulars of which they are predicated.) What's more, if the
one is not an independent thing, it is clear that neither could number
have being as some nature separated from things (for number consists
of units, while the unit is that of which the very being is oneness). But
if there is something which itself is and is one, the thinghood of it must
be oneness and being, since there is not any other thing as a result of
which they are attributed but just they themselves.

1001a 30   But surely if there should be some being-itself and one-itself, there
is a considerable impasse about how there would be anything else
besides these—I mean how beings will be more than one. For what
is other than being *is* not, so that, according to the argument of

1001b   Parmenides, it necessarily follows that all things are one and that this
is being. But there is trouble both ways, for whether the one is not
an independent thing or there is some one-itself, it is impossible that
number be an independent thing. Why this is so has been said above,
if there is no one-itself; but if there is, the same impasse results as
about being. For from what source will there be another *one* besides
the one-itself? What's more, if the one-itself is indivisible, then by a
principle of Zeno's it would be nothing. For that which, when added
to something does not make it greater, and when subtracted from it

does not make it less, he says is not a being, obviously taking a being
to be a magnitude; and if it is a magnitude it is one of a bodily sort,
since this has being in every direction. With the other magnitudes,
being added to in one way will make them greater, but in another
way will not, as with a surface or a line, but with a point or an
arithmetic unit, in no way will it do so. But since he is looking at
things in a crude way, and it is possible for there to be something
indivisible, therefore in this way it has a rebuttal even against him.
For such a thing when added to something will not make it greater,
but will make there be more of them. But how indeed could there be
magnitude out of one such thing or more than one of them? For it
would be like claiming the line is made of points. But surely even if
one conceives it in such a way that, as some say, number has come
into being out of the one itself and something else that is not one,
nonetheless one must inquire why and how the thing that comes into
being will at one time be a number but at another time be a magnitude,
if the thing other than one is to be the unequal, and be the same
nature. For it is not clear how magnitude could have come into being
either from the one and this nature, or from some number and this
nature.

Chapter 5 An impasse following upon these is whether numbers,
solids, surfaces, and points are or are not independent things. For if
they are not, what being is escapes us, as well as what the thinghood
of beings is; for attributes, motions, relations, dispositions, and ratios
seem not at all to signify thinghood (since all of them are attributed
to some underlying thing, and none of them is a *this*). But the things
that would seem most of all to signify thinghood, namely water, earth,
fire, and air, out of which composite bodies are put together, have
hotness, coldness, and such things as attributes, not as thinghood,
while only the body that is subject to these persists and is a being
and an independent thing. But surely the extended body is less an
independent thing than is the surface, and that less so than is a line,
and that than a unit or a point; for by these the body is limited, and
while they seem to be possible without a body, without them the body
is impossible. For this very reason, even though most people and the
earlier thinkers supposed being and thinghood to be body, with the
other things attributes of this, so that the sources of bodies would be
the sources of beings, the thinkers who came later and seem to be wiser

than these others take the sources to be numbers. So as we are saying, if these are not independent things, nothing at all is an independent thing, nor is anything a being, since the attributes of these are not worthy of being called beings.

But now if this is agreed to, that lengths and points are thinghood more than bodies are, though we do not see what sort of bodies these belong to (since it is impossible that they be among perceptible things), there could not be any independent thing at all. What's more, all these 1002a 20 seem to be divisions of bodies, in breadth, depth, or length. And on top of this, any shape at all is equally present in the solid, or none is; so if Hermes is not in the stone, neither is the half of the cube in the cube in the manner of a determinate thing.[10] Therefore the surface is not in it either (since if any surface at all were, so would be this one that delineates the half of it), and the same argument applies also to the line, the point and the unit; so if as much as one wishes the body is an independent thing, but these things more so than it, while it is not possible that these are particular independent things, then what being is and what the thinghood of beings is escape us.

In addition to the things that have been said, what concerns coming 1002a 30 into being and destruction also turns out to be irrational. For it seems that an independent thing, if it now *is* after not being before, or later *is not* after having been before, undergoes these things in a way that involves coming into being or passing away. But points, lines, and surfaces do not admit of either coming into being of passing away, even though they sometimes are and sometimes are not. For whenever 1002b extended bodies touch or are divided, they at once become one when they touch, and at once become two when they are divided. Therefore when they are put together, something is not but has been destroyed, and when they are divided there are things which previously were not (for the point, which is indivisible, has surely not been divided in two), and if things come into being and are destroyed, out of what do they come into being? It is just about the same way as with the now in time; for neither does this admit of coming into being and passing away, but still there always seems to be a different one, inasmuch as it is not an independent thing. So it is clear that it is the same way too

---

10 This apparent conclusion is denied at the end of Bk. V, Ch. 7, and more fully refuted in Bk. IX, Ch. 6–7.

with points, lines, and surfaces; for the argument is the same, since all   1002b 10
of them alike are either boundaries or divisions.

Chapter 6 One might be at an impasse as to why at all one has to
look for any other things besides perceptible ones and the in-between
ones, such as those we posit as forms. If it is because the mathematical
things, while they differ from the things around us in some other way,
do not differ from them at all to the extent that there are many of them
of the same kind, so that the sources of them would not be limited
in number (just as the sources of all the written things around us are
not limited in number either, but in kind, unless one takes the sources
of some particular syllable or utterance, of which they will be limited   1002b 20
even in number, it is likewise with the in-between things, since there
too the things of the same kind are infinite), so that, if there are not,
apart from the perceptible things and mathematical things, some other
things of the sort that some people say the forms are, there would not
be any independent thing that is one in number as well as in kind,
nor would the sources of things be of a certain definite number except
in kind—if, then, this is necessary, it is also necessary on this account
that there be forms. For even if those who say there are forms do not
articulate the reason very well, still this is what they mean, and it is
necessary for them to say that each of the forms is an independent
thing and none of them is an attribute. But if we do set down that there   1002b 30
are forms and that the sources of things are one in number and not in
kind, we have spoken of the impossible things that must follow.[11]

In the same area as these things is questioning whether the elements
have being potentially or in some other way. For if it is in some other
way, something else would be more primary than the sources (for the   1003a
potency is more primary than some particular cause, and everything
that has the potency need not be some particular way). But if the
elements do have being potentially, it is a possibility that none of the
beings would have being. For even what is not yet the case is capable
of being so, since what is not the case does come to be so, but none of
the things that are incapable of being so come to pass.

So it is necessary to raise both these impasses about the sources,
and one as to whether they are universal or what we call particular.

---

[11] See 999b 32–1000a 4.

1003a 10

For if they are universals, they will not be independent things. (For none of the common predicates signifies a *this* but rather an of-this-sort, while an independent thing is a this; while if the thing predicated in common *were* a this and were to be set out apart, Socrates would be many animals—himself as well as human being and animal—if each of them signifies something that is one and a this.) So if the sources are universal, these thing follow; but if they are not universal but *are* in the same way as particulars, there will be no knowledge, since of all things the knowledge is universal. Therefore, if there were going to be knowledge of them, there would be other sources prior to the sources, predicated of them in a universal way.

Book IV (Book Γ)
The Study of Being as Being[1]

Chapter 1 There is a kind of knowledge that contemplates what is 1003a 21 insofar as it is, and what belongs to it in its own right. And this is not the same as any of those that are spoken of as partial, since none of the other kinds of knowledge examines universally what pertains to being as being,[2] but cutting off some part of it, they consider this attribute, as do the mathematical kinds of knowledge. But since we are seeking the sources and the highest causes, it is clear that they must belong to some nature in its own right. So if also those who were seeking the elements of beings were in quest of these sources, the elements too must belong to what is, not incidentally but insofar as it is. For this 1003a 30 reason, for us too it is the first causes of being as being that must be gotten hold of.

Chapter 2 Being is meant in more than one way, but pointing toward one meaning and some one nature rather than ambiguously.[3] And just as every healthful thing points toward health, one thing by protecting it, another by producing it, another by being a sign of health, and another because it is receptive of it, 1003b and also what is medical points toward the medical art (for one thing is called medical by having the medical art, another by being well suited to it, another by being an action belonging to the

---

[1] This title for Book IV supplied by the translator.

[2] This is one of two ways that Aristotle describes the topic of first philosophy, or metaphysics. He also calls it the study of the highest kind of being, which is separate and motionless. (1026a 17–18) Many commentators find these accounts incompatible. The latter makes metaphysics a theology, but the study of being as such, they say, would be an ontology, seeking the common structure or lowest common denominator of all beings. This and the following chapter explain, however, that being as such belongs only to the highest kind of being. Only it could have the attribute of being in its own right, rather than incidentally or derivatively. (See the Introduction.)

[3] Aristotle has argued, beginning at 998b 22, that being is not a class that includes all beings, but neither is it simply an ambiguous word like the English "bark." While the meanings of being are irreducibly different, they are all governed by one primary meaning, just as, in the following example, there would be no such thing as a healthy diet or a healthy blood sample if there were not, in the primary sense of the word, a healthy animal.

medical art, and we shall find other things spoken of in a similar way to these), so too is *being* meant in more than one way,
but all of them pointing toward one source. For some things are
called beings because they are independent things, others because
they are attributes of independent things, others because they are
ways into thinghood, or destructions or deprivations or qualities of
thinghood, or are productive or generative of independent things,
or of things spoken of in relation to independent things, or nega-
1003b 10   tions of any of these or of thinghood, on account of which we
even say that nonbeing *is* nonbeing. So just as there is one kind
of knowledge of all healthful things, this is similarly the case with
the other things as well. For it is not only about things meant in
one way that it belongs to one kind of knowledge to contemplate
them, but also about things meant in ways that point toward one
nature, for these too are in a certain manner meant in one way.
Therefore it is clear that it belongs to one kind of knowledge also
to contemplate beings as beings. And knowledge is always chiefly
about what is first, on which other things depend and through
which they are named. So if this is thinghood, the philosopher
would need to understand the sources and causes of independent things.

　　Now of every class that is one there is both one perception and
1003b 20   one kind of knowledge, as the grammatical art, which is one, considers all utterances; and for this reason it also belongs to a kind
of knowledge that is generically one to study as many forms as
there are of being as being, and to the species of it to study their
species. But if being and oneness are one and the same nature in
that they follow upon one another, as do source and cause, though
not as being revealed in a single articulation (though it makes no
difference even if we do understand them alike, but is even more
convenient)—for "one human being" and "a human being that *is*"
and "a human being" are the same thing, and nothing different is
revealed by the redoubled statement that the human being is one
or that the human being that is one *is,* and it is clear that they
1003b 30   are not distinct with respect to either coming into being or passing
away, and similarly with oneness, so that obviously the addition in
these statements signifies the same thing, and what is one is nothing different aside from what is, while further the thinghood of each
thing is not incidental but is likewise the very thing that something

is[4]—therefore, just as many forms as there are of oneness, so many also are there of being. It belongs to a kind of knowledge that is generically the same to consider what pertains to what these are; I mean, for example, about sameness and similarity and other such things. And just about all opposites lead back to this starting point; but let these have been examined by us in the passages about contraries.[5]

1004a

And there are as many parts of philosophy as there are kinds of thinghood, so that there must necessarily be among them some part that is primary and some part that follows upon it. For being starts right out already having classes, on account of which the kinds of knowledge also follow along with these. For the philosopher is described just as the mathematician is, since this study too has parts, and there is a certain first and second kind of knowledge, and others in sequence among mathematical things.

But since it belongs to one study to examine opposites, while manyness is opposite to oneness, it belongs to one study to examine negation and deprivation since what is examined in both cases is the one thing of which there is a negation or deprivation. For we either say simply that that one thing is not present, or that it is not present in some class; in the latter case a difference is attached to the one thing aside from what is in the negation, since the negation is the absence of it, but in the deprivation some underlying nature comes along to which the deprivation is attributed. Therefore also the things opposite to the ones mentioned, otherness and dissimilarity and inequality, and as many other things as are spoken of as consequences of these, or of manyness and oneness, belong to the kind of knowledge mentioned for their being known. And among these is also oppositeness, since oppositeness is a certain kind of difference and difference a kind of otherness. So since oneness is meant in more than one way, these too will be meant in more than one way, even though it belongs to one study to know them all. For it is not when something is meant in more than one way that the kinds of knowledge are different, but when the

1004a 10

1004a 20

---

[4] An atomist would deny both of Aristotle's claims here, asserting not only that a being such as an animal is really many, but that its thinghood is an incidental result of collisions and adhesions.

[5] The reference is unknown. There is an extended discussion of the topics mentioned here in Bk. X, Ch. 3–9.

meanings neither refer back directly to a single one nor point toward a single one. But since they do all refer back by pointing toward a primary meaning, as whatever is called one points toward the primary unity, one must likewise say that this holds also concerning sameness, difference, and contraries; so having distinguished in how many ways each of them is meant, in this way one must give an account of how each one is intended to point toward that which is the primary instance of each attribute. For some point toward it by having it, others by making it, and still others will be meant according to other such ways. It is clear, then, that it belongs to one kind of knowledge to have a reasoned account about these things and about thinghood (and this was one of the things contained among the impasses).

1004a 30

1004b   And it belongs to the philosopher to be capable of considering all things. For if it does not belong to the philosopher, who will it be who examines whether Socrates and Socrates sitting down are the same, or whether one thing is contrary to one thing, or what a contrary is, or in how many ways it is meant? And it is the same with the other such things. So since these things are attributes of oneness as oneness and of being as being in their own right, but not insofar as things are numbers or lines or fire, it is clear that it belongs to that kind of knowledge to know what they are as well as the attributes of them. And those who do examine these topics go wrong not in the sense that they are not philosophic, but because thinghood, to which they pay no attention, is prior. Seeing that, just as there are also attributes proper to number as number (such as oddness, evenness, commensurability, equality, being greater, and being less) and these belong to numbers both on their own and in relation to one another (and similarly there are other attributes proper to what is solid and either motionless or moving, either weightless or having weight), so too to being as being certain attributes are proper, it is these about which it belongs to the philosopher to investigate the truth. Here is a sign of this: those who engage in dialectic and the sophists slip into the same outward appearance as the philosopher. For sophistry is wisdom in appearance only, while dialectic discourses about everything, and being is common to all things, so it is clear that they discourse about these topics just because they are proper to philosophy, for sophistry and dialectic turn themselves to the same class of things as philosophy, but it differs from one of them in the way its power is turned, and from the other in the choice of a way of life it makes; dialectic is tentative

1004b 10

1004b 20

about those things that philosophy seeks to know, and sophistry is a seeming without a being.

Further, one of the two rows of corresponding contraries is a list of deprivations, and they all lead back to being and nonbeing, or to oneness and manyness, as, for example, rest belongs to oneness and motion to manyness, and almost everyone agrees that beings and thinghood are put together out of contraries. At least they all say that the sources are contraries, since some say they are the odd and the even, others the hot and the cold, others the limit and the unlimited, and others friendship and strife. And all the other things obviously lead back to oneness and manyness (for let the derivation be granted by us), while the sources received from the other thinkers also fall wholly within these classes. So it is clear from these things too that it belongs to one kind of knowledge to study being as being. For all things either are contraries or are derived from contraries, and the origins of the contraries are oneness and manyness. But these belong to one kind of knowledge, whether they are meant in one sense, or even if they are not, as presumably the truth has it. But notwithstanding that oneness is meant in more than one way, the other ways will be meant to point toward the primary meaning, as will, similarly, the contraries, even if being or oneness is not a universal that is the same in every instance nor anything separate, as presumably they are not, but rather some meanings of them point toward a single one, while others are in a series successively derived.

For this reason it does not belong to the geometer to study what contrariety or completeness or oneness or being or sameness or difference are, except as based on a hypothesis. That, then, it belongs to one kind of knowledge to study being as being and the things that belong to it insofar as it is being, is clear, and that the same contemplative study is about not only independent things but also what belongs to them, the ones mentioned as well as what is prior and posterior, genus and species, whole and part, and the other such things.

Chapter 3 One must discuss whether it belongs to one kind of knowledge or to different ones to be concerned with the things that are called axioms in mathematics as well as with thinghood. And it is evident that the inquiry about these things belongs to one kind of knowledge and that this is the one that belongs to the philosopher; for they belong to all beings and not to some particular class separate

1004b 30

1005a

1005a 10

1005a 20

from the rest. And everyone uses them, because they belong to being as being, and each class of things is a class of beings; but people use them only so far as is sufficient for them, and this is as far as the class of things extends about which they are carrying out demonstrations. So since it is clear that they belong to all things insofar as they are beings (since this is what is common to them), the study of the one who knows about being as being is also about these things. For this reason,

1005a 30   no one who is engaged in particular kinds of inquiry says anything about them, whether they are true or not, neither the geometer nor the arithmetician, though some of those who study nature do, and do so appropriately; for they suppose that they alone inquire about the whole of nature and about being. But since there is something still higher than what is natural (for nature is one particular class of being), the inquiry about these axioms would belong to the contemplative

1005b   study that is universal and directed toward the primary kind of being. The study of nature is a kind of wisdom too, but not the primary one. However much some of those who speak about truth try to say in what way one must receive it, they do this through a lack of education in the arts of logical analysis; for one must arrive already knowing about these things, and not find them out while studying.[6]

So it is clear that it belongs to the philosopher and the one who studies all thinghood, insofar as it is by nature, to investigate also about the starting points of demonstrative reasoning; and it is appropriate for the one who most of all knows each class of things to be able to

1005b 10   state the most certain sources of what he is concerned with, so that the one who knows about being as being would be able to state the most certain sources of all things. And this is the philosopher. And the most certain of all principles is that about which it is impossible to be in error; for such a principle must be the best known (since all people are deceived about things they do not know) and nonhypothetical. For that which is necessary for one who understands any of the beings whatever to have is not a hypothesis; and that which is necessary for one who knows anything whatever to know is necessary for him to arrive having.

---

[6] In Bk. II, Ch. 3, this same point is made in the context of whether mathematical precision is appropriate in all kinds of argument. Here it has the effect of setting the study of logic outside of and preparatory to the properly philosophic studies, among which metaphysics is primary and physics secondary.

That, then, such a principle is the most certain of all, is obvious; what it is, after this prelude, let, us state. It is not possible for the same thing at the same time both to belong and not belong to the same thing in the same respect[7] (and as many other things as we ought to specify in addition for the sake of logical difficulties, let them have been specified in addition[8]). And this is the most certain of all principles, since it has the distinction mentioned. For it is impossible for anyone at all to conceive the same thing to be and not be, as some people think Heracleitus says.[9] For it is not necessary that someone take what he says into his understanding. And if contraries cannot belong at the same time to the same thing (and let the usual things be specified in addition for us in this proposition too), and an opinion is contrary to the opinion that contradicts it, it is clear that it is impossible for the same person at one time to believe the same thing to be and not be. For one who is in error about this would have contrary opinions at the same time. For this reason everyone who demonstrates traces things back to this as an ultimate opinion, since this is by nature a source even of all the other axioms.

1005b 20

1005b 30

Chapter 4 There are some people who, as we say, themselves claim that it is possible for the same thing to be and not be, and also claim that it is possible to conceive something this way. And in fact many of the writers about nature use this way of speaking. But we have just taken it as understood that it is impossible to be and not be at the same time, and by means of this we have shown that this is the most certain of all principles. Yet some people expect even this to be demonstrated, but on account of lack of education, for it is a lack of education not to know of what one ought to seek a demonstration and of what one ought not. For it is impossible that there be a demonstration of absolutely everything (since one would go on to infinity, so that not even so would there be

1006a

---

[7] The axioms about which this chapter speaks are sometimes called "laws of thought." This formulation already makes it clear that Aristotle considers them principles that govern being rather than thinking.

[8] Examples of these hairsplitting difficulties are given by Socrates in Plato's *Republic*, 436B–437A.

[9] See especially fragments 32, 51, and 62 in the Diels numbering. Heracleitus uses the tension of naked and unexplained contradictions to shock the hearer out of superficial habits of thought. See also 1010a 10–16 and note.

1006a 10   a demonstration), and if there are certain things of which one ought
             not to seek a demonstration, these people are not able to say what they
             think would be of that kind more than would such a principle.

             But even about this there are ways to demonstrate that it is impos-
             sible by means of refutation, if only the one disputing it says something;
             if he says nothing, it is absurd to seek an argument to meet someone
             who has no argument, insofar as he has none, for such a person, insofar
             as he is such, is from that point on like a plant. I say that demonstrat-
             ing by means of refutation is different from demonstrating because the
             one demonstrating would seem to require from the outset the thing
             to be shown, while if someone else is responsible for such a require-
             ment, there would be a refutation rather than a demonstration. And
             the starting point for all such arguments is not the demand that one
1006a 20   say something either to be or not to be (for perhaps one might suppose
             that this would require from the outset the thing to be shown), but
             that what he says must *mean* something to both himself and someone
             else; for this is necessary, if he is going to say anything. For if this is
             not the case, there would be no argument with such a person, neither
             by himself in relation to himself, nor with anyone else. But if someone
             grants this, there will be a demonstration, for there will already be
             something determinate. But the one responsible for it is not the one
             demonstrating but the one who submits to it, for while doing away
             with reason he submits to reason. Furthermore, the one who concedes
             this has conceded that something is true without demonstration.[10]

             In the first place, then, it is clear that this very thing is true, that
1006a 30   the words *be* or *not be* mean something definite, so that not everything
             could be so and not so. Further, if "human being" signifies one thing,
             let this be "two-footed animal."[11] Now by signifying one thing, I mean

---

[10] This beautiful passage sometimes overshadows the rest of the chapter. One might
think that by choosing not to speak, one could avoid the necessity of refraining
from self-contradiction. But without such a principle, one could not even have any
experience, as the rest of the chapter shows. Agreeing to speak is only required to
make the necessity of the principle of contradiction manifest, without obliging anyone
to assume that principle from the outset.

[11] Since Aristotle elsewhere gives better definitions of a human being, that are just
as brief, why does he use this one here? He does sometimes use nonsense definitions
(e. g., at 1029b 27–8) to keep a line of argument distinct, and this one does recall
Academic jokes about "featherless bipeds," but there is a subtle and serious point
underlying it. Aristotle has just compared the one who refuses to speak to a plant,
and later in this chapter (1008b 14–18) argues from footedness to choices to implicit

this: if this is a human being, then if anything is a human being, this will be its being-human. (And it makes no difference if one says it means more than one thing, but only a limited number, since one could set down a different word for each formulation; I mean, for instance, if one says that "human being" means not one thing but many, of which the definition of one would be "two-footed animal," while there would also be a number of others, but limited in number, then one could set down a special name for each of the definitions. But if one were not to posit this, but said it meant infinitely many things, it is clear that there would be no definition; for not to mean one thing is to mean nothing, and when words have no meaning, conversation with one another, and in truth, even with oneself, is abolished. For it is not possible to think without thinking one thing, so if thinking is possible, one could set down one name for this thing.) So let there be, as was said at first, some meaning for the name, and one meaning. Now it is not possible that being-human should mean just exactly *not* being-human, if "human being" not only signifies something belonging to one thing but also one meaning. (For we do not regard what belongs to one thing as having one meaning, since in that way even educated, white, and human would mean one thing, so that all things would be one, since they would be synonyms.)

And it will not be possible to be and not be the same thing other than ambiguously, as in the case that what we call "human being," other people were to call "not human being"; but the thing raising an impasse is not this, whether it is possible for the same thing at the same time to be and not be a human being in name, but in respect to the thing. Now if "human being" and "not human being" mean nothing different, it is obvious that not being human will be nothing different from being human, so that being human would *be* not being human, since they would be one. For this is what it means to be one thing, as are a robe and a cloak, that the meaning of the definition is one; and if they are to be one, being human and not being human must mean one thing. But it was shown that they mean different things. So it is necessary, if it is true to say that something is a human being, that it be a two-footed animal (since this was what "human being" was to

1006b

1006b 10

1006b 20

1006b 30

philosophic opinions. The necessary sequence from two-footed animal to rational animal is traced in Erwin Straus's splendid essay "The Upright Posture," in his book *Phenomenological Psychology* (Basic Books, 1966).

mean); but if this is necessary, it is not possible for the same thing not to be a two-footed animal (for this is what it means to be necessary, that it be impossible for something not to be). Therefore it is not possible for it to be true at the same time to say that the same thing is and is not a human being.

1007a   And the same argument also applies to not being a human being. For to be a human and not to be human mean different things, if to be white and to be human are different; for the former is much more opposed to being human than the latter is, and therefore means something different. And if someone is going to say that "white" means one and the same thing as "human being," we in turn will say just the same thing that was said before, that all things would be one, and not only opposites. But if this is not possible, then what was said follows, if the person being questioned gives an answer. But if, when the question is

1007a 10   asked simply, he also gives denials, he is not answering the question. For nothing prevents the same thing from being human and white and a countless multitude of other things; but nevertheless, upon being asked if it is true to say that this is a human being or not, one must answer the one thing meant and not give the extra answers that it is also white and big. For it is not even possible to go through all the attributes, since they are infinite, so then let them all be gone through or none. So likewise, even if the same thing is human and not human ten thousand times over, one ought not to give an extra answer to the question whether it is a human being, saying that it is at the same time also not a human being, unless one must also answer in addition as many other things as incidentally go along with it, both as many as it

1007a 20   is and as many as it is not; but if one does this, there is no conversation.

In general, those who say this do away with thinghood and what it is for something to be. For it is necessary for them to say that all things are incidental, and that there is not anything which is the very thing it is to be human or to be an animal. For if there is to be anything which is the very thing it is to be a human being, this will not be not being-human or being not-human (in fact these are negations of it); for there was one thing that it meant, and this was the thinghood of some particular thing. But to signify thinghood is to mean that nothing else is the being of it. But if the very thing that it is to be human is the same as the very thing that it is either not to be human or to be not human,

1007a 30   then something else *would* be the being of it, so that they would have to say that nothing has any such defining articulation, but all things are incidental; for in this respect thinghood and what is incidental are distinguished. For whiteness is incidental to a human being because, even if he is white, white is not the very thing that he is.

But if all things are meant as incidental, there will be no first thing about which they are meant, if the incidental always signifies a predication about some underlying subject. Therefore it must go   1007b
on to infinity. But that is impossible, since no more than two things are intertwined; for what is incidental is not incidental to something incidental, unless it is because both are incidental to the same thing. I mean, for instance, that the white thing is educated and this is white because both are incidental to a human being. But it is not the case with the educated Socrates that both things are incidental to something else. So since some things are spoken of as incidental in the latter way and others in the former, all those that are meant in the latter way, in which whiteness is incidental to Socrates, do not admit of being infinite on the higher side, such that something else could be incidental to the   1007b 10
white Socrates; for no one thing comes into being out of all of them. But neither would anything else be incidental to whiteness, such as being-educated; for this is no more incidental to that than that is to this, and at the same time it was distinguished that some things are incidental in this way but others in the way that being educated is incidental to Socrates. For all those of the latter kind, the incidental is not incidental to the incidental, but rather for those of the former kind, so not everything will have been meant as incidental. Therefore there will also be something signifying thinghood. But if this is so, it has been shown that it is impossible for contradictory things to be attributed at the same time.[12]

What's more, if contradictory things are all true of the same thing at the same time, it is obvious that all things will be one. For the same   1007b 20
thing would be a battleship and a wall and a human being, if something admits of being affirmed or denied of everything, as it must be for those

---

[12] The "law of contradiction" has now been taken a step deeper than the way it first came to sight. Even if the world consisted of random collections of changing attributes, nothing could be and not be the same thing at the same time and in the same respect. But thinghood brings with it enduring unity. Something is or is not a human being simply, for as long as it is at all, and the two qualifications (same time and same respect) need not be added.

who repeat the saying of Protagoras.[13] For if the human being seems to someone not to be a battleship, it is clear that he is not a battleship; and so he also is one, if the contradictory is true. And so the claim of Anaxagoras comes true, that all things are mixed together, so that nothing is truly any one thing. They seem, then, to be talking about the indeterminate, and though supposing they are talking about what is, they are talking about what is not; for the indeterminate is that which has being potentially, and not in full activity. But surely they are

1007b 30   obliged to state either the assertion or the denial of everything about everything; for it is strange if its own denial belongs to each thing, while the denial of something else that does not belong to it would not belong to it. I mean, for example, if it is true to say that a human being is not a human being, it is clear that he also either is a battleship or is not a battleship. Then if the assertion can belong, then necessarily the denial can too; while if the assertion does not belong, then the denial

1008a   at any rate would belong more than would the thing's own denial. So if even that denial belongs, the denial of the battleship would also belong, and if that can, then so can its assertion.

These things follow, then, for those who make this argument, and it also follows that it is not necessary either to assert or to deny something. For if it is true that it is a human being and is not a human being, it is clear that it will also be *neither* a human being nor a nonhuman being. For there are two denials for the two statements, and if the former pair makes one claim composed of both, the latter would also be one claim opposite to it.

Further, either things are this way about everything, and it is both white and not white, and something that is and something that is not,

1008a 10   and assertions and denials in a similar way apply to the rest, or this is not so but applies to some things and not to others. And if it does not apply to everything, these exceptions would be agreed to; while if it does apply to everything, again either to all things that the assertion applies, the denial does as well, and to all things that the denial applies, the assertion does as well, *or* to those to which the assertion applies, the denial does too, while to those to which the denial applies, the assertion does not apply to all. And if it is this way, there will be something that

---

[13] "A human being is the measure of all things—of the things that are, that they are, and of the things that are not, that they are not."

is unqualifiedly not the case, and this would be an established opinion; and if what is not the case is something established and known, the opposite affirmation would be more known. But if it is likewise also necessary to assert everything that is denied, it must either be true to state them separately, for example that something is white and in turn that it is not white, or not. And if it is not true to state them separately, one is not even saying these things, and in fact one is not anything (for how could nonbeings utter a sound or walk?), and all things would be one, just as was said before, and the same thing will be a human being and a god and a battleship and the contradictories of these. (For if it is similar with each thing, there will be no difference of one from another; for if something were to differ, that would be true and peculiar to it.) And likewise, even if something admits of being true separately, what was said follows, and in addition to this, that everyone would be right and everyone would be wrong, and one agrees that one is wrong. But at the same time it is clear that the investigation on the part of this person is about nothing, since he says nothing. For he says neither that it *is* this way or that it is not this way, but that it both is and is not this way, and then in turn he states the denial of both, and says that it neither is nor is not this way, for if that were not so, something would already be determinate.

Yet if whenever the assertion is true, the denial is false, and if this is true, the assertion is false, it would not be possible truly to assert and deny the same thing at the same time. But perhaps one might say that this is what was assumed at the outset.

Yet can it be that the one who conceives that something either is or is not a certain way is in the wrong, while the one who conceives it both ways is in the right? For if he is in the right, what would it mean that the nature of beings is of this kind? But if he is not in the right, but more in the right than the one who conceived things in the former way, beings would already be a certain way, and this would be true, and not at the same time also not true. But if everyone alike is both in the wrong and speaks the truth, it will not be possible for such a person to utter or say anything; for at the same time he says both these things and not these things. But if one conceives nothing, but alike believes and does not believe, how would he be in any different condition from plants? And from this most of all it is obvious that no one is in this condition, neither anyone else nor those stating this argument. For why does he walk to Megara and not sit still, when he thinks he ought to walk? And

1008a 20

1008a 30

1008b

1008b 10

why does he not march straight into a well in the morning, or straight over a cliff, if it happens that way, but obviously take care, as though not believing that falling was both good and not good? Therefore it is clear that he does conceive of one thing as better and the other as not better. But if this is so, it is necessary that he also conceive of one 1008b 20   thing as a human being and another as not a human being, and one thing as sweet and another as not sweet. For he does not seek after and conceive of all things equally when, supposing that it is better to drink water or see some human beings, he thereupon seeks after them; and yet he would have to, if the same thing were alike both a human being and not a human being. But, just as was said, there is no one who does not obviously take care about some things and not about others; therefore, as it seems, everyone conceives things to be simply a certain way, if not about everything, at least about what is better and worse. And if they do so not knowing but having opinion, one ought to be much more concerned about the truth, just as one who is sick ought to 1008b 30   be more concerned about health than one who is healthy. For one who has an opinion, as compared with the one who has knowledge, is not disposed in a healthy way toward truth.

On top of this, if all things are so and not so as much as one could wish, still at least the more and the less are present in the nature of things. For we would not say that two and three are even in just the same way, nor that the one who thinks four things are five is just as wrong as the one who thinks they are a thousand. So if they are not equally wrong, it is clear that one of the two is less wrong and therefore 1009a   more right. Then if what is more is nearer, there would be something true to which what is more right is nearer. And even if there is not, still there is already something more stable and more trustworthy, and we would be set free from the anarchic argument that prevents anything from being made definite in our thinking.

Chapter 5 As a consequence of the same opinion, there is also the pronouncement of Protagoras,[14] and they must both alike either be so or not be so; for if all opinions and appearances are true, it is necessary 1009a 10   that all things be at the same time both true and false (for many people conceive of things that are contrary to one another, and consider those

---

[14] See note to 1007b 23.

who do not believe the same things as themselves to be wrong, so that the same thing must both be and not be), and if this is so, all opinions must be true (for people who are wrong and those who are right believe things opposite to one another, so if things are this way, everyone is right). So it is clear that both of these arguments are consequences of the same thinking. But the same way of going about things is not possible in every encounter; for some people need persuasion while others need brute force. For all those who believe this way from being at an impasse in thought, their ignorance is easy to cure (for the engagement is not with their argument but with their thinking); but for those who argue for the sake of arguing, the remedy for them is a refutation of their exact vocal and verbal argument.

1009a 20

For those who are at an impasse, this opinion has arisen from perceptible things, that contradictories and contraries are present at the same time, since they see opposites come into being out of the same thing. So if what is not cannot come to be, the thing was from the beginning both ways alike, just as Anaxagoras says everything is mixed in everything, and Democritus too. For he says that both the void and the full are alike present in any part of things whatever, even though one of these is as being and the other as nonbeing. To those who believe on these grounds, then, we shall say that in a certain way they speak rightly, but in another way they go wrong. For being is meant in two ways, so that there is a way in which it is possible for something to come into being out of what is not, and a way in which it is not, and the same thing at the same time can both be and not be, but not in the same respect. For it is possible for the same thing at the same time to be contrary things potentially, but not in full activity. And what's more, we shall expect them to understand that there is among beings a certain other kind of independent thing to which there belongs neither motion nor destruction nor any becoming at all.

1009a 30

And similarly, for some people the truth about appearances has come out of perceptible things. For they think it is not appropriate that what is true be judged by manyness or fewness, while the same thing seems to some of those who taste it so be sweet but to others to be bitter, so that if everyone were sick or everyone were insane while two or three people were well or had sense, these latter would seem to be sick or insane and the others would not. Further, opposite things appear to be the case about the same things to many of the animals and to us, and to each person himself as compared with himself, the same

1009b

things do not always appear to be so as a result of sense perception.

1009b 10   Which sort among these are true or false is not clear; for the ones are no more true than the others, but just like them. For this reason, in fact, Democritus says that either nothing is true or at any rate things are obscure to us.

But in general, because of assuming that knowledge is sense perception, and that this in turn is a process of being altered, people say that the appearance that comes from sensing is necessarily true; for it is from these reasons that both Empedocles and Democritus, as well as, so to speak, each of the others has become vulnerable to opinions of this sort. For Empedocles says that those who change their condition change their knowledge, "for wisdom grows for humans according to

1009b 20   what is present." And in other verses he says,

However much change comes into their natures, just that much
Does it always happen to them to think changed thoughts.

And Parmenides declares himself in the same way:

For as at any time is the composition of the much-twisted limbs,
Thus is intellect present to humans; for it is the very same thing
That thinks, the nature of the limbs both in all humans and in
Each one, since its thought is what is more in its mixture.

And a blunt remark of Anaxagoras to some of his friends is also remembered, that beings would be for them however they conceive them. And people also say that Homer seemed to have this opinion,

1009b 30   because he made Hector,[15] when he was knocked out by a blow, lie "thinking changed thoughts," as though even those who are delirious are thinking, just not the same things. So it is clear that, if both are processes of thinking, the beings too are at the same time so and not so. And it is in this respect that what follows is most harsh: for if those who most of all have seen the truth that is accessible—and these are those who seek it most and love it most—if *they* have such opinions and declare these things about truth, how will this not be enough to make those who are trying to philosophize lose heart? For seeking the

1010a   truth would be a wild goose chase.

Now the cause of their opinion is that they were inquiring into the truth about beings, but they assumed that the only beings were perceptible things; but among these the nature of the indeterminate

---

[15] Actually Epeius (*Iliad* XXIII, 689).

is heavily present, and the sort of being that we described. For this reason, while they speak reasonably, they do not speak the truth (for it fits better to say it this way than the way Epicharmus spoke of Xenophanes[16]). And further, it was because of seeing all nature around us in motion, while about what is changing nothing is true, or at least does not admit of being true about what is wholly changing in every way. For out of this conception the most extreme opinion of those   1010a 10 mentioned burst into bloom, that of the people who announce that they are Heracleiteans,[17] and of the sort that Cratylus held, who at last believed that it was necessary to say nothing but only moved his finger, and who censured Heracleitus for saying that it is not possible to step into the same river twice, since *he* believed it is not possible even to step into it once.

Now we too will say in response to this argument that there is some reason for them to believe that what is changing, when it is changing, does not have being. It is, however, something disputable, since a thing that is losing something has some of what it is losing, and a thing must already be something of what it is becoming. And in general, if   1010a 20 something is being destroyed, there will be present something that *is*, and if something is coming into being, there must be something out of which it is coming into being and something by which it is generated, and this cannot go on to infinity. But passing over those things, let us say these: that to change in quantity is not the same as to change what sort of thing something is, so letting something not stay constant with respect to size, we still recognize everything on account of its form. What's more, those who conceive things this way deserve to be censured because, having seen that things are in this condition with a lesser number even of the perceptible things themselves, they declared that it is the same way for the whole heaven; for the place around us in the sensible world is the only one that is constantly in a state of passing away and coming to be, but this is, so to speak, not even a piece of the   1010a 30 whole, so that it would have been more just if they had acquitted these

---

[16] Apparently, that he spoke truly but unreasonably.

[17] See 1005b 24–26 and note. Heracleitus was most famous for such paradoxes as "everything is in flux" and the one about the river that is quoted below, to which those who adopted his name were attracted. In fact, however, the central idea in his writings is that of the *logos* which brings stability out of the midst of flux, a thought very close to Aristotle's own idea of being-at-work.

things on account of those instead of condemning those on account of these. And further, it is obvious that we shall also say to them the same things that were said earlier; for that there is some motionless nature must be shown to them and they must be convinced of it. And in fact it follows for those who say everything is and is not at the same time that they should say that everything is at rest rather than in motion; for there is not anything into which anything can change, since everything
1010b   belongs to everything.

But as far as truth is concerned, that not every appearance is true is because, in the first place, even if the perception, at least of the proper object of each of the senses,[18] is not false, still appearance is not the same thing as perception. Then too, it is worth wondering at if they are at an impasse about this: whether magnitudes are of the amount or colors of the sort that they appear to be to those far away or those up close, or whether they are the way they appear to the healthy or to the sick, and whether things are heavier that seem so the weak or to the strong, and whether those things are true that seem so to those who
1010b 10   are asleep or those who are awake. That they do not really believe it is obvious; at any rate, no one, even if he thinks he is in Athens one night when he is in Libya, goes to the Athenian town hall. And about the future too, as Plato says,[19] surely the opinion of the doctor and that of an ignorant person are not alike authoritative about, say, who is or is not going to get well. Further, among the senses themselves, the perceptions of an alien or a proper attribute are not alike authoritative, nor are those by a similar sense and the proper sense itself,[20] but about color, sight is authoritative, not taste, and about flavor, taste is, not

---

[18] See *On the Soul* 418a 8–17. This view is a reversal of Democritus's famous saying that all immediately perceived qualities, such as sweet, are only conventions. Modern thinkers such as Galileo, Descartes, and Locke dismiss them again as merely "subjective," sometimes calling them secondary qualities. In the twentieth century, Husserl and others again reassert the primacy of sensory experience.

[19] *Theaetetus,* 177C–179B. This passage is far-reaching, establishing the good as neither relative nor subjective. Aristotle too, in the *Nicomachean Ethics* (1134b 24–35), says that there is a natural justice, easily distinguished from customs about right and wrong.

[20] Examples of the two preceding clauses might be these: a lemon looks both bitter and yellow, and both smells and tastes bitter. The second judgement of the eye, and the second judgement of the bitterness, are the ones we give more credence to.

sight, each of which at the same time about the same thing never says
that it is simultaneously so and not so. But not even at a different time
does one of the senses disagree about the *attribute,* but only about that
to which the attribute belongs. I mean, for example, that the same wine
might seem at one time to be sweet and at another time not to be, if
either it or one's body had changed; but the sweetness itself, such as
it is whenever it might be present, never changes, and one is always
right about it, and what is going to be sweet is necessarily of such a
kind. And yet all these arguments abolish this, and just as they make
thinghood be nothing, so too do they make nothing be necessary; for
what is necessary cannot be otherwise and otherwise again, so that if
anything is so by necessity, it will not be both so and not so.

And in general, if all there is is what is perceptible, nothing would
be if there were not beings with souls, since there would be no sense
perception. Now that there would be neither sense objects nor sense
perceptions is perhaps true (for these are experiences of the perceiver),
but it is impossible that there not be the underlying things which
bring about sense perception, even without the perception. For the
perception itself is surely not *of* itself, but there is also something else
besides the perception, which must be prior to the perception; for what
causes motion is prior by nature to what is moved, and even if these
things are meant relatively to one another, this is nonetheless so.[21]

Chapter 6 There are some of those who are convinced of these
things, as well as some who merely repeat these arguments, who raise
an impasse: they inquire who it is who judges which person is healthy,
and generally who judges each thing correctly. But such perplexities
are like being in doubt whether we are now asleep or awake, and all
such impasses amount to the same thing. For these people insist that
there be a justification of everything; for they ask for a starting point,
and want to get hold of it by demonstration, when the fact that they are
not convinced of their own doubts is obvious in their actions. Instead,
the very thing we are saying is the condition they are in, for they are

---

[21] In general, relative terms name things that depend on each other and come into
being together (such as double and half, master and slave, etc.), but knowledge and
perception are exceptions, being relative to the knowable and perceivable which are
not relative to them. See *Categories* 7b 15–8a 12.

looking for a reason for what has no reason, since the starting point of
demonstration is not a demonstration. These people, then, could easily
be persuaded of this (since it is not hard to understand), but those who
are looking only for the brute force of an argument are looking for what
is impossible, for they think it is alright to say contradictory things,
and to go right on saying them.

But if it is not the case that everything is relative, but some things are
themselves in their own right, then not everything that appears would
be true. For what appears is an appearance to someone, so that the one
1011a 20   who says that all appearances are true makes all beings relative. And
for this reason, those who are looking for the brute force of argument,
and at the same time insist on putting forth a reason, ought to beware,
since it is not the appearance that is true but the appearance to the one
it appears to, and when it appears and in what respect and how. And
if they do give an account but do not give it this way, it will soon turn
out that they say contradictory things. For it is possible for something
to appear to the same person to be honey by sight but not by taste,
and, since there are two eyes, not even to appear the same to each
of them by sight, if they are unlike. So as for those who say, for the
1011a 30   reasons mentioned above, that what appears is true, and hence that
all things are alike false and true, since it is not the case either that the
same things appear to everyone or the same things always to the same
person, but often contrary things appear at the same time (for touch
says there are two things between our crossed fingers while sight says
there is one), nevertheless something does not appear in contrary ways
1011b   to the same sense in the same respect in the same way *and* at the same
time, so that *this* would be true. But perhaps for this reason those who
argue not on account of an impasse but for the sake of arguing need to
say not that this is true but that it is true for this person. And, as was
said before, they must make all things relative and dependent upon
opinion and perception, so that nothing either has happened or will
be so if no one has thought so first. But if anything *has* happened or
will be so, it is clear that all things could not be relative to opinion.
What's more, if something is one, it is related either to one thing or to a
definite number of things; and if the same thing is both half and equal,
1011b 10   still it is not in relation to its double that it is equal. But if the same
thing is a human being and thinkable thing in relation to a thinker,
not the thinker but only the thinkable thing could be a human being.

And if each thing is relative to the one thinking, the thinker would be relative to things infinite in kind.[22]

So to the effect that the best established of all opinions is that contradictory statements are not true at the same time, and what follows for those who say that they are, and why they talk that way, let so much have been said. But since it is impossible for a contradiction to be *true* at one time of the same thing, it is clear that neither could contrary properties *belong* to the same thing at the same time. For it is no less so that one of the two contraries is a deprivation, and a deprivation of thinghood, while a deprivation is a negation of some definite class. So if it is impossible to assert and deny truly, it is also impossible for contraries to be present together, unless both are present in certain respects, or one of them in a certain respect, and the other simply.

1011b 20

Chapter 7 But neither is it possible that there be anything between contradictories, but about any one thing whatever, it is necessary either to affirm or deny one of them. This is evident first of all to those who define what the true and the false are. For to say that what is is not or that what is not is, is false, but to say that what is is and what is not is not, is true,[23] so that the one who says that something is or is not is either right or wrong. But if there is a middle ground, neither what is nor what is not is said either to be or not to be. What's more, the thing between the contradictories would be between them either as gray is between black and white, or as what is neither one is between human being and horse. Now if it is between in this latter way, it could not change (for something changes from, say, not-good to good, or from that to not-good), but as it is, in-between things always seem to be involved in change (for there is no change other than into opposites or what is in-between). On the other hand, if it is in-between in the

1011b 30

---

22 To say that *everything* is relative to human opinion would mean that there is no such thing even as human opinion, in two ways: it would itself only be what it is in relation to human opinion, and it would dissolve into an infinity of relations. The first consequence means that there is nothing to be on the other side of the relation to opinion, and the second means there is no one thing to be part of each distinct relation.

23 This is the first of three progressively deeper definitions of truth in the *Metaphysics*. The others are in Bk. VI, Ch. 4 and Bk. IX, Ch. 10.

1012a   former way, in this way too there would be something that now is
not seen: some turning into white that was not from something not-
white. Again, our thinking either affirms or denies everything that it
thinks about—and this is clear from its definition—whenever it thinks
truly or falsely; whenever it puts together an assertion or denial this
way, it is true, but that way, it is false. Further, there would have to
be in-between things alongside all contradictories, if one is not just
arguing for the sake of arguing, and therefore there will be something
that is neither true nor not-true, and something besides being and not-
being, as that there would be a kind of change that is not becoming or
passing away. Again, in all those cases of things in which the negation

1012a 10   of something implies its contrary, there will be in-between things even
among these, as among numbers, a number that is neither odd or not-
odd, which is impossible, as is obvious from the definition. Further,
this will go on to infinity, and beings will not only be half again as many
but more. For it will be possible in turn to negate both the affirmation
and the denial of this in-between thing, and this will be something,
for its thinghood will be something other than the previous one. And
further, whenever one who is asked if a thing is white says that it is
not, he has negated nothing other than its being so, while the not being
so is the negation.

Now this opinion has come about for some people in the same
way that other paradoxes have; for whenever one is not able to refute

1012a 20   a debater's arguments, by giving in to the argument, he concedes
the thing on which the reasoning was based to be true. Some people,
then, say this for some such reason, but others because of seeking an
argument for everything. But the starting point for all of these comes
from definition. And the definition arises from the necessity that they
mean something, since the articulation of which the word is a sign
will be a definition. But while the Heracleitean account, saying that
everything is and is not, seemed to make all things be true, that of
Anaxagoras, that there is something between contradictories, seems
to make all things false, for when things are mixed, the mixture is
neither good nor not-good, so that there is nothing true to say.

Chapter 8 Now that these things have been distinguished, it is clear

1012a 30   the statements that go only one way about all things cannot be accepted
in the way some people say them, some of them saying that nothing is
true (for they say that nothing prevents everything from being just like

"the diagonal is commensurable"[24]), while others say that everything is true. For these claims are just about the same as the Heracleitean one; for whoever says that all things are true *and* all things are false also makes each of these two claims separately, so that if they are both impossible, it too is impossible. Yet there are obviously contradictory things that cannot be true at the same time; nor indeed can all things be false, even though this might seem more possible from what has been said. But in the face of all such claims one must require what was said in the chapters above, not that something be or not be so but that it *mean* something, so that one must reason from a definition to grasp what would mean something false or something true. And if what is true to assert is what is false to deny, it is impossible for all things to be false, since one of the two parts of the contradiction must be true. Further, if it is necessary either to assert or deny everything, it is impossible that both be false, since *one* of the two parts of the contradiction is false. And in fact all these claims are subject to the much repeated point that they annihilate themselves. For the one who says that everything is true also makes the statement contrary to his own be true, so that his own is not true (since the contrary one says that it is not true), while the one who says that everything is false also himself makes himself wrong. And if they make exceptions, the former that the contrary of his claim is the only one that is not true, or the latter that his own is the only one that is not false, nevertheless it turns out that they require infinitely many statements to be false or true, since the one that says that the true statement is true is also true, and this goes on to infinity.

It is also clear that neither those who say that all things are at rest or those who say they are all in motion speak the truth. For if all things are at rest, the same things would always be true and the same things false, but this obviously changes (for the one who says this himself at one time was not and again will not be); but if all things are in motion, nothing would be true and therefore everything would be false, but it has been demonstrated that this is impossible. What's more it must be what is that changes, since change is from something to something. But it is surely not everything that is at rest sometimes and in motion

1012b

1012b 10

1012b 20

1012b 30

---

[24] It *seems* perfectly obvious that some small fractional part of a square's diagonal must fit some exact number of times into its side, but precise reasoning shows that this is impossible. The sophist counts on undermining one's trust in whatever seems evident.

sometimes, while nothing is always at rest or in motion, for there is something that is always moving the things that move, and the first mover itself is motionless.

Book V (Book Δ)
Things Meant in More than One Way[1]

Chapter 1 *Source* means that part of a thing from which one might first move, as of a line or a road there is a source in one direction, and another one in the opposite direction; and it means that from which each thing might best come into being, as in the case of learning, sometimes one ought to begin not from what is first and the source of the thing, but from which one might learn most easily; or it means that constituent from which something first comes into being, such as the keel of a ship or the foundation of a house,[2] and in animals some say it is the heart, others the brain, and others whatever they happen to believe is of this sort; or it means that which is not a constituent, from which something first comes into being, and from which its motion and change naturally first begin, as a child from its father and mother, or a fight from insults; or it means that by whose choice what is moved is moved or what changes changes, in the sense in which the ruling offices of cities as well as oligarchies, monarchies, and tyrannies are called sources,[3] as are the arts, and among these the master crafts most of all. Also, that from which a thing is first known is called the source of the thing, such as the hypotheses of demonstrations.

Causes are meant in just as many ways, since all causes are sources. And what is common to all sources is to be the first thing from which something is or comes to be or is known; of these, some are present within while others are outside. For this reason nature is a source, as are elements, thinking, choice, thinghood, and that for the sake of which;

1012b 34
1013a

1013a 10

1013a 20

---

[1] This title for Book V supplied by the translator. The secondary literature generally refers to this book as a dictionary, but Aristotle himself, at the beginnings of Books VII and X, says that it is about the various ways things are meant. The point is not to define words but to collect and organize the distinct senses of important words meant in more than one way. These ambiguities are not verbal but inherent in things, and Aristotle steadfastly preserves them.

[2] Of Odysseus's house, the source in this sense is his marriage bed, rooted in the earth (*Odyssey* Book XX, lines 190–201).

[3] There is a faint echo of this meaning when we speak of going to the source for a decision, but it is the predominant sense of the Greek word; *archē* means not only what was first but what is first, a ruling beginning.

for the good and the beautiful are sources of both the knowledge and
the motion of many things.

Chapter 2   *Cause* means, in one sense, that out of which something
comes into being, still being present in it, as the bronze of a statue or the
silver of a bowl, or the kinds of these. In another way it is the form or
pattern, and this is the gathering in speech of what it is for something
to be, or again the kinds of this (as of the octave, the two-to-one ratio,
or generally number), and the parts that are in its articulation. In yet
1013a 30   another it is that from which the first beginning of change or rest is,[4]
as the legislator is a cause, or the father of a child, or generally the
maker of what is made, or whatever makes a changing thing change.
And in still another way it is meant as the end. This is that for the
sake of which,[5] as health is of walking around. Why is he walking
around? We say "in order to be healthy," and in so saying think we
have completely given the cause. Causes also are as many things as
1013b   come between the mover of something else and the end, as, of health,
fasting or purging or drugs or instruments. For all these are for the sake
of the end, but they differ from one another in that some are deeds and
others tools.

The causes then are meant in just about this many ways, and
it happens, since they are meant in more than one way, that the
nonincidental causes of the same thing are also many (as of the statue
both the art of sculpture and bronze, not as a consequence of anything
else but just as a statue, though not in the same way, but the one
as material and the other as that from which the motion was). And
1013b 10   there are also causes of one another (as hard work of good condition
and this in turn of hard work, though again not in the same way,
but the one as end and the other as source of motion). Further, the

---

[4] This is sometimes mistakenly called the efficient cause. Aristotle never describes it in
such a way, and we generally intend by the phrase the proximate cause, the last event
that issues in the effect. Aristotle always means instead the *origin* of motion, when it
happens to be outside the moving thing. It is only in a derivative or incidental sense
that he will speak of a push or a bump as being a cause at all, since, as he says at
1013a 16 above, all causes are sources.

[5] This is Aristotle's comprehensive phrase for what is generally called the final cause,
a much broader idea than that of purpose. The *telos,* or completion, of a deliberate
action or series of actions is a purpose, but in nature the completion for the sake of
which things occur is the wholeness of natural beings.

same thing is a cause of opposite things, for the thing that is present is responsible for this result, and we sometimes blame this, when it is absent, for the opposite one, as the absence of the pilot for the ship's overturning, whose presence was the cause of its keeping safe, and both, the presence and its lack, are causes as movers.

But all the causes now being spoken of fall into four most evident ways. For the letters of syllables and the material of processed things and fire and earth and all such things, of bodies, and parts of a whole and hypotheses of a conclusion are causes as that out of which, and while the one member of each of these pairs is a cause as what underlies, such as parts, the other is so as the what-it-is-for-it-to-be, a whole or composite or form. But the semen and the doctor and the legislator, and generally the maker, are all causes as that from which the source of change or rest is, but other things as the end or the good of the remaining ones. For that-for-the-sake-of-which means to be the best thing and the end of the other things, and let it make no difference to say the good itself or the apparent good.

1013b 20

The causes then are these and so many in form, but the ways the causes work are many in number, though even these are fewer if they are brought under headings. For cause is meant in many ways, and of those of the same form, as preceding and following one another. For example, the cause of health is the doctor and also the skilled knower, and of the octave the double and also number, and always things comprehensive of whatever is among the particulars. Further, there is what is incidental, and the kinds of these, as of the statue, in one way Polycleitus and in another the sculptor, because it is incidental to the sculptor to be Polycleitus. And there are things comprehensive of the incidental cause, as that a human being is the cause of the statue, or generally an animal, because Polycleitus is a human being and a human being is an animal. And also among incidental things, some are more remote and others nearer, as if the pale man or the one with a refined education were said to be the cause of a statue. And all of them, both those meant properly and those incidentally, are meant some as potential and others as at-work, as of building a house, either the builder or the builder building. And similarly to the things that have been said, an account will be given for those things of which the causes are causes, as of this statue or a statue or in general an image, and of this bronze or of bronze or in general of material, and likewise with the incidental things. Further, things intertwining the former and

1013b 30

1014a

1014a 10

the latter will be said, such as not Polycleitus nor a sculptor but the sculptor Polycleitus.

Nevertheless, all these are six in multitude, but spoken of in a twofold way: there is the particular or the kind, the incidental or the kind of the incidental thing, and these intertwined or spoken of simply, and all as either at-work or in potency. And they differ to this extent, that what is at-work and particular is and is not present at the same time as that of which it is the cause, as this one healing with this one being cured or this one building with this thing being built, but not always so with what is potential. The house and the housebuilder are not finished off simultaneously.

1014a 20

Chapter 3 *Element* means that out of which something is composed, as the first constituent not divisible in kind into a different kind, as the elements of speech are those things out of which speech is composed and into which it is last divided, while they are no longer divided into other utterances different in kind from them, but if they are divided, the parts are of the same kind, as a part of water is water, though a part of a syllable is not a syllable. Similarly, the elements of bodies are meant by those who say into what bodies are divided last, while those things are no longer divided into others differing in kind; and whether there is one such thing or more than one, they call them elements. And the elements of geometrical proofs are meant in much the same way, and so are the elements of demonstration in general, for the first demonstrations are also present in the rest of the demonstrations, and these are called elements of demonstration; and of this kind are the primary syllogisms, made of three terms and concluded through one middle term.

1014a 30

1014b

Also, by altering the sense from this one, people call what is one and small, and useful for many things, an element, and so anything small, simple, and indivisible is called an element. Hence it comes about that the most universal things are elements, because each of them, being one and simple, is present in many things, either all or as many as possible, and that oneness and the point seem to some people to be sources of things. So since the so-called genera[6] are universal and

1014b 10

---

[6] Aristotle tends to confine the word genus to the kinds that are one step above the species of natural beings, each being the primary constituent in the articulation of what something is, as he puts it below at 1024b 4–5. The use of the word referred

indivisible (since there is no articulation of them), some people call the genera elements, and more so than the specific difference because the genus is more universal; for that to which the specific difference belongs, the genus follows along with, while the specific difference does not belong to everything that the genus does. But common to all the meanings is that to be an element in each sense is to be a first constituent in each thing.

Chapter 4   In one sense, *nature* means the coming into being of things that are born, as if one had said "nativity,"[7] but in a sense it means the first thing present in a growing thing, from which it grows. Again, it is that from which the primary motion in each of the beings that are by nature is present within itself insofar as it is itself. And those things are said to grow which have increase from something else by contact and either growing-into-one or, as with embryos, growing-upon, while growing-into-one differs from contact in that there nothing else besides the contact is necessary, but in things that grow into one there is some one thing that is the same in both which makes them grow together instead of touching, and be one by continuity and quantity, though not by what sort of thing each is.

1014b 20

And further, nature means the primary thing from which any of the beings that are not by nature is or comes to be, which is unarranged and unable to be changed out of its own capacity, as bronze is said to be the nature of a statue or of bronze implements, or wood of wooden things, and similarly in other cases; for each of them is made out of these things because the primary material is preserved throughout. In this way too the elements of the things that are by nature are said to be nature, some people saying that it is fire, some earth, some air, some water, some some other such thing, others some of these, and still others all of them. And again, in another way nature means the thinghood of things that are by nature, as those people mean who say that nature is the primary combination of things, or as Empedocles says

1014b 30

1015a

No nature belongs to any of the things that are;

---

to here presupposes pressing merely logical definitions back to an array of highest classes of all things.

[7] Literally, "as if one were to pronounce the upsilon in *phusis* long."

There is only mixture and remixture of intermingled things, and
Nature is given as a name by human beings.

For this reason, though all things that are or come to be by nature
already have present that out of which they naturally come into being
or are made, we say that they do not yet have their natures if they do
not have their look and form. So that which consists of both of these
is by nature, such as animals and their parts, and nature is both the
first material (and this in two senses, either first in relation to the thing
itself or first generally, as with works of bronze, the bronze is first in
1015a 10   relation to them, but in general it is perhaps water, if all things that
melt are water) and the form or thinghood, which is the completion of
a thing's coming into being. And in an extended sense, every kind of
thinghood in general is directly called a nature for this reason: because
nature is a certain kind of thinghood.

So from what has been said, the primary and authoritative meaning
of nature is the thinghood of things that have in themselves a source
of motion in their own right; for the material is called nature by being
receptive of this, and coming-into-being and growing are called nature
by being motions proceeding out of this. And this is the source of
motion of things that are by nature, being present in them all along in
some way, either potentially or fully at work.

1015a 20   **Chapter 5**   *Necessary* means that without which, as a contributing
cause, it is not possible to live, as breathing and food are necessary
to an animal, since without them it is impossible for them to be, and
it means those things without which the good either could not be
or could not come about, or the bad be cast off or avoided, as, say,
drinking medicine is necessary in order not to be sick, or sailing to
Aegina in order to get money. Also, it refers to what is compelled, or
to force, and this is what constrains or prevents, contrary to impulse
and choice, for what is forced is called necessary, and for this reason
necessity is also painful, as Evenus says "For every necessary thing is
1015a 30   by nature annoying," and force is a kind of necessity as Sophocles says
"But force made it necessary for me to do this"; and necessity seems
to be something that cannot be changed by persuasion, and seems so
rightly, for it is contrary to motion that results from choice and from
reasoning.

Further, we say that which is incapable of being otherwise to be
as it is necessarily, and it is as a result of *this* necessity that all other

necessities are in some way attributed; for what is forced means what is necessary to do or suffer whenever, on account of being forced, one is incapable of acting from impulse, as though this kind of necessity were one through which something could not be otherwise, and it is similar in the case of contributing causes of life or of the good. For whenever without something anything is incapable in one case of the good, or in the other case of life and being, these things are necessary and this cause is a kind of necessity. Also, demonstration is of necessary things, because it cannot be otherwise, if it has simply been demonstrated, and the causes of this are the first premises, if those from which the conclusion comes cannot possibly be otherwise. So of some things, something else is the cause of their being necessary, but of others, nothing else is, but rather it is through them that other things are necessary. Therefore the primary and authoritative sense of necessary belongs to what is simple, for this is not capable of being in more than one way, and so has no this way and otherwise, for then it would automatically have more than one way of being. So if there are some things that are everlasting and motionless, for them nothing is forced or contrary to nature.

Chapter 6   *One* is meant in one sense of what is so incidentally, in another sense of what is so in its own right; incidentally one, for example, are Coriscus and educated, or educated-Coriscus (for it is the same thing to say Coriscus and educated, and educated-Coriscus), or educated and just, or educated and just Coriscus. For all these are called one incidentally, the just and the educated because they are incidental to one independent thing, and educated and Coriscus because one of them is incidental to the other; and similarly, in a certain way educated-Coriscus is one with Coriscus, because one of the parts in the expression is incidental to the other one, namely educated to Coriscus, and educated-Coriscus is incidental to just-Coriscus because one part of each expression is incidental to the same single thing. And it is similar even if the incidental thing is attributed to a class of things, or to phrases which name some universal, as if one were to say that human being and educated human being are the same; for this is either because educated is incidental to human being, while it is one independent thing, or because both of them do not belong to him in the same way, but the one presumably as a class and in his thinghood, while the other is a state or attribute of the independent thing.

1015b

1015b 10

1015b 20

1015b 30

As many things, then, as are called one incidentally are meant in this way, but of those called one in their own right, some are meant as being continuous, as a bundle is held together by means of a cord, and wood by means of glue; and a line, even if it is bent, so long as it is continuous, is called one, just as also each of the body parts is, such as a leg or an arm. Among these themselves, the things that are continuous by nature are one more so than those that are continuous by means of art. And what is called continuous is that of which the motion is one in its own right,[8] and not capable of being otherwise, while the motion is one if it is undivided and in an undivided time. And those things are continuous in their own right which are one not by contact; for if you place pieces of wood touching each other, you would not say that these are one either as wood or as a body, nor that they are continuous in any other way. And things that are continuous are in general called one, even if they have a bend, and still more so those that do not have a bend, as the shin or thigh is one more than is the leg, because it is possible for the motion of the leg not to be one. And the straight line is one more so than the bent line, but the line that is bent and has an angle we call both one and not one, because it is possible for its motion both to be and not be simultaneous; but the motion of the straight line is always simultaneous, and of no part of it that has magnitude does part stay still while part moves, as do parts of the bent line.

Also in another way, things are called one because what underlies them is undifferentiated in kind; and those things are undifferentiated whose form is not divisible into subclasses by sense perception, while what underlies them is either first or last from the end. For wine is called one, and water is called one, in that they are indivisible in kind, but also all juices, such as olive oil and wine, as well as all things that melt, are called one because the last thing underlying them all is the same, since they are all water or air. But also those things are called one whose genus is one, even though they differ by opposite

1016a

1016a 10

1016a 20

---

[8] The meanings of *one* in this chapter are arranged in ascending order, from weakest to strongest. Two of the steps on the way are weaker and stronger senses of continuity (in Greek, the property of holding together). This clause is not a general definition of continuity, but a description of a sufficient sign of its weakest sense. If you pick up one of the sticks tied into a bundle, the rest come with it. A necessary condition of the stronger sense of the word is given in the *Physics* at 232b 24–25, and a proper definition of it at 227a 11–12.

specific differences, and these are all called one because the genus that underlies their differences is one (for example, a horse, a human being, and a dog are one thing because they are all animals), in much the same way as the material is one, And while these things are sometimes called one in this way, sometimes it is said that the higher genus is the same, if they are ultimate species of the genus, as the isosceles and equilateral are one and the same figure, since they are both triangles, but not the same triangles.[9]   1016a 30

Again, all those things are called one of which the articulation saying what it is for them to be is indivisible into any other one revealing what the thing is. (Every articulation itself is divisible within itself.) For in this way, even what is growing or shrinking is one thing, because its articulation is one, just as in the case of the articulation of the shape of plane figures. And in general, those things of which the *thinking* is indivisible, which thinks what it is for them to be, if it is not capable of separating them in time or in place or in articulation, are one most of all, and of these, most of all those which are independent things. For generally, whatever does not have a division, insofar as it does not have it, is in that respect called one; for example, if insofar as it is a human being it has no division, it is one human being, if insofar as it is an animal, it is one animal, or if insofar as it is a magnitude, it is one magnitude. So while most are called one by way of something else, either doing or having or undergoing or being in a relation to something that is one, others are called one in the primary sense, and of these the thinghood is one, and one either in continuity or in species or in articulation, for we count as more than one those things that are not continuous, or of which the species is not one, or of which the articulation is not one. But again, there is a sense in which we say that anything whatever is one if it is so-much and continuous, but there is another sense in which we do not if it not some kind of whole, that is, if it does not have a form that is one; for instance, we could not say that it was one all the same if we saw the parts of a shoe put together any which way, unless on account of continuity, but only if they were put together in such a way that it *is* a shoe and already has some one   1016b   1016b 10

---

[9] In this way, instead of saying that the horse and dog are one thing because they are both animals, one would say they are one and the same life form, animals and not plants, though not the same animal.

form. And for this reason among lines, the shape of the circle is one most of all, because it is whole and complete.

To be one is to be a source for something to be a number; for the first measure is a source, since that by which we first know each class of things is the first measure of it. So oneness is the source of what is knowable about each thing. But what is one is not the same in all classes; for here it is the smallest musical interval, but there it is the vowel or consonant, and of weight it is a different thing, and of motion something else. But what is one is always indivisible, either in amount or in kind. As for what is indivisible in amount, that which is indivisible in any way and has no position is called the arithmetic unit, that which is indivisible in any way and has position the point; that which is divisible in one direction is a line, in two a plane, and what is divisible in all three directions with respect to amount is a bodily solid, and—going back the other way again—what is divisible in two directions is a plane, in one a line, what is divisible in no direction with respect to amount is a point or unit, the one without position the unit, the one with position a point.

Again, some things are one in number, others in species, others in genus, and others by analogy: in number, things of which the material is one, in species things of which the articulation is one, in genus, things to which the same manner of predication applies,[10] and by analogy, as many things as are in the condition that something else is, in relation to something else. The later ones follow along with the earlier ones, as things that are one in number are also one in species, but not all those that are one in species are one in number; but as many things as are one in species are all also one in genus, while those that are one in genus are not all one in species, but are all one by analogy, but not all those that are one by analogy are one in genus.

And it is clear that *many* will be meant in ways opposite to one: for some things are many through not being continuous, others through having their material divided in kind, whether the first or last material, and others because the articulations that say what it is for them to be are more than one.

Chapter 7 *Being* is meant in one sense incidentally, in another sense

1016b 20

1016b 30

1017a

---

10 In this sense, all qualities are one, all places, etc. See below, beginning at 1017a 23.

in its own right; in the incidental sense, we say, for example, that the just
person is educated, or the human being is educated, or the educated
one is a human being, in much the same way as if we were to say that the   1017a 10
educated one builds a house because it is incidental to the housebuilder
to be educated, or to the educated one to be a housebuilder (for here
this *is* this means that this is incidental to this). And it is this way
too in the case of the things mentioned; for whenever we say that the
human being is educated or the educated one is a human being, or that
the white thing is educated or this is white, we mean in some cases
that both are incidental to the same thing, in others that something is
incidental to a being, and in the case of the educated human being, that
the educated is incidental to this person. (And in this sense even the
not-white is said to "be" because that to which it is incidental *is*.) So
things that are said to be incidentally are said to be so either because   1017a 20
both belong to the same being, or because one of them belongs to a
being, or because the thing itself is, to which belongs that to which it
is attributed.

But just as many things are said to be in their own right as are meant
by the modes of predication; for in as many ways as these are said, in so
many ways does *to be* have meaning. Since, then, of things predicated,
some signify what a thing is, others of what sort it is, others how much
it is, others to what it is related, others what it is doing or having done
to it, others where it is, and others when it is, being means the same
thing as each one of these.[11] For it makes no difference whether one
says a person is healing or a person heals, or a person is walking or
cutting rather than that a person walks or cuts, and similarly in the   1017a 30
other cases.

Also, *to be* and *is* signify that something is true, and *not to be* signifies
that it is not true but false, alike in the cases of affirmation and denial;
for instance, that Socrates *is* educated indicates that this is true, or
that Socrates *is* indicates that this is true, but that the diagonal *is not*
commensurable indicates that this is false.

Again, *being* and *what is* mean in one sense something that is definite   1017b
as a potency, but in another sense what is fully at work, among these

---

[11] These modes of predication, or ways of saying anything about anything, are also the
ultimate classes of beings, not able to be reduced in number. They are usually referred
to as the "categories." The book called the *Categories,* of which the authorship is
sometimes disputed, adds two more to the eight given here.

things that have been mentioned. For we say of both one who is capable of seeing and one who is fully at work seeing that he sees, and similarly of both one who is capable of using knowledge and one who is using it that he knows, and also both of that to which rest already belongs and that which is capable of being at rest that it rests. And it is similar in the case of independent things, for we say that Hermes is in the block of stone, and that the half belongs to a line, and that what is not yet ripe is grain. When something is potential and when it is not yet so must be distinguished in other places.[12]

1017b 10   **Chapter 8** *Thinghood* is attributed to the simple bodies, such as earth, fire, water, and whatever is of this sort, and also to bodies in general and the things composed of them, both living things and heavenly bodies as well as their parts; and all these are called independent things because they are not attributed to anything underlying them, but other things are attributed to them. But in another way, thinghood means that which is responsible for the being of a thing, and is a constituent in whatever things are of such a kind as not to be attributed to an underlying thing; an example is the soul of an animal. Further, thinghood refers to whatever parts are present in such things that mark them off and indicate a *this,* the removal of which does away with the whole; as a body is annihilated by the removal of its surface, as some

1017b 20   say, or a surface by the removal of its boundary line; and in general, number seems to some people to be this sort of thing (since nothing would be if it were removed, and it marks off all things). But it also means what it is for something to be, the articulation of which is a definition, and this is called the thinghood of each thing.

It turns out, then, that thinghood is meant in two ways, both as the ultimate underlying thing, which is no longer attributed to anything else, and also of whatever is a *this* and separate, and of this sort is the form or "look" of each thing.[13]

---

[12] See Book IX, Ch. 7.

[13] Commentators are quick to deny what Aristotle says here, citing other places where he says that the form is *not* separate except in thought. But this comes from a failure to understand the dialectical structure of Aristotle's writings, which follow the order of inquiry. For instance, it is said in this chapter and again at the beginning of Bk. VII, Ch. 2, that the parts of animals are independent things because, at first glance, their organs and systems seem isolable. But at the beginning of Bk. VII, Ch. 16, as the inquiry into thinghood nears its conclusion, Aristotle rejects that preliminary opinion,

Chapter 9 Things are said to be the *same* in some instances incidentally, as what is white and what is educated are the same because they are incidental to the same thing, or a human being and educated because one of them is incidental to the other, and the educated thing is human because it is incidental to a human being. And this combination is the same as either part, and either of them is the same as it, for both the human being and the educated one are said to be the same as the educated human being, and it to be the same as they. (And for this reason all these statements are made nonuniversally, since it is not true to say that every human being is the same as the educated; for things that are universal belong to things in their own right, while incidental things belong to them not in their own right. But they are predicated of particulars simply; for Socrates and educated Socrates seem to be the same thing, but "Socrates" is not applied to many things, on which account "every Socrates" is not said in the way that "every human being" is.)

1017b 30

1018a

But while some things are said to be the same in this way, others are said to be so in their own right, and in exactly as many ways as they are said to be one; for those things of which the material is one, either in kind or in number, are also said to be the same, as well as those of which the thinghood is one, so it is evident that sameness is a kind of oneness of the being of things that are either more than one, or that are being used as more than one, such as when one says that a thing itself is the same as itself, since one is using it as two. But things are called *other* of which either the forms, the material, or the articulation of the thinghood is more than one, and in general other is meant in ways opposite to same.

1018a 10

All those things are called *different* that are other but are the same in some respect, only not in number but in either species or genus or by analogy. Also called different are those things of which the genus is other, and contraries, and as many things as have otherness in their thinghood.

---

and explains why. In the same way, it must seem to begin with that forms have a separate being only in our thinking, and Aristotle says this at various places, but at 1072a 25–28, the culmination of the *Metaphysics* and of all Aristotle's philosophizing, he deduces the existence of separate forms. The present sentence is an anticipation of that inquiry, and means exactly what it says.

Alike is used of things that have the same attributes in every respect, as well as of things that have more attributes the same than different, and of which the quality is one; and in the case of contrary attributes which are capable of altering, a thing is like that one which shares either the most or most important of these. And *unlike* is meant is ways opposite to like.

1018a 20   **Chapter 10** Things said to be opposites are contradictories, contraries, relative terms, lacking and having, and the extremes from which things come into being and into which they pass away; and whatever things do not admit of being present at the same time in something that is receptive of both of them are said to be opposed, either themselves or what they are made of. For gray and white do not belong to the same thing at the same time because what they are made of is opposed.

    Contraries mean things not capable of being present in the same thing at the same time, which differ in genus, or things in the same genus that differ most, or things in the same recipient that differ most, 1018a 30   or things that come under the same capacity that differ most, or things whose difference is greatest either simply, or in a genus or species. The other things that are called contraries are so called either because of having such things, or because of being receptive of them, or because of being productive of or affected by them, or being losses or gains or states of having or lacking them. And since one and being are meant in more than one way, all those other things that are meant in ways corresponding to these must follow along with them, so that also same, other, and contrary are different according to each of the ways of predicating being.

1018b   Different in species are all those things which are of the same genus and are not subordinated one to the other, and all those things which are in the same genus and have a difference, and all those things that have a contrariness in their thinghood; contraries are also different from one another in species, either all of them or the ones called so in the primary sense, as are all those things in the ultimate species of a genus that have different articulations (as human being and horse are undivided in genus, but have different articulations), as well as all those things which, though present in the same independent thing, have a difference. The same in species are those things that are meant in ways opposite to these.

Chapter 11   Things are said to be *preceding* and *following* in some cases, if there is some first thing or beginning in each class, because of being nearer to some beginning determined either simply and by nature, or else relatively or somewhere or by some people, as some things precede others in place by being nearer either to some definite natural place (such as a middle or an end), or to some random thing, while what is farther from it is following. Other things are preceding in time (some by being farther from the present, in the case of past things, for the Trojan wars precede the Persian ones because they stand farther away from the present, but others by being nearer to the present, in the case of future things, for the Nemean games precede the Pythian ones because they are nearer, when one uses the present as a beginning and first time). Other things have precedence in motion (for what is nearer to its first mover precedes the rest, as a boy precedes a man, while an origin of motion takes precedence simply). Other things have precedence in power (for what exceeds in power takes precedence and is more powerful; and of this sort is the person whose choice someone else must go along with and follow, so that the latter does not move when the former does not set him in motion, and does move when he does, the choice being the origin). Other things precede in an ordering (and these are all those things that are placed at intervals in relation to one definite thing in accordance with some pattern, as the second member of a chorus precedes the third, or the middle note of a chord precedes the highest, for there the beginning is the choral leader, and here it is the tonal center).

In this sense, then, these things are called preceding, but in another sense what precedes in knowledge take precedence simply. And among these, that which has precedence according to reason does so in a different way from that which does so according to sense perception, for according to reason it is the universal that takes precedence, but according to sense perception it is the particular; and according to reason the attribute precedes the whole, as the educated precedes the educated human being, since there could not be the whole articulation without the part, and yet it is not possible for the educated to be if there is not something that is educated. Also, the attributes of things that have precedence are said to have precedence, such as straightness over smoothness, since the former is an attribute of a line on its own, but the latter of a surface.

Now some things are called preceding and following in that way,

1018b 10

1018b 20

1018b 30

1019a

but others in accordance with nature and thinghood, namely those things that are capable of *being* without other things, while those others are not capable of being without them, which is a distinction that Plato used. (But since being is of more than one kind, first of all the underlying subject takes precedence, on account of which the independent thing precedes, but secondly precedence varies in accordance with potency or being-at-work; for some things precede in potency but others in being-at-work. In potency the half line precedes the whole one, the part precedes the whole, and material precedes

1019a 10

the independent thing, but in being-at-work they follow them, for only when the thing is broken apart will they be fully at work.) So in a certain way all the things that are said to precede and follow are meant in accordance with this distinction; for some things are capable of being without others as a result of coming into being, such as the whole without the parts, but others as a result of destruction, such as the part without the whole. And it is similar with the rest.

Chapter 12 *Potency* means a source of motion or of change, either in something else or in something *as* something else; for example, housebuilding is a potency that is not present in the thing that is built, but medical skill is a potency which could be in the one who is healed, but not insofar as he is healed. So the source of change or of motion in

1019a 20

something else or as something else is called a potency, but so is the source of being moved or changed by something else or as something else. For in virtue of that by which something passive is acted upon, we say that it is capable of being acted upon, sometimes in any way whatever, but sometimes not with respect to every affection but only for the better. It also means the power of carrying this out well or in accordance with choice; for sometimes of those who just travel or speak, but not well or not in the way they intend, we say that they are not capable of speaking or walking, and similarly in the case of being passive. Also, all those states in virtue of which things are completely unaffected or unchangeable, or else not easily altered for the worse, are called potencies; for things are broken and crushed and bent and

1019a 30

in general destroyed not by being capable but by being incapable and falling short of something; while among such things, those are unaffected which hardly and barely can be acted upon because of a potency and by being capable and holding on in some state.

And since potency is meant in so many ways, also what is capable

will mean in one way what has a source of motion or change (for
even what can bring something to a stop is something capable) in
something else or as something else, and in one way even if some
other thing has such a power over it, and in one way even if it has a
power to change in any direction whatever, whether for the worse or
for the better. (For even what is destroyed seems to be capable of being
destroyed, since it would not be destroyed if it were incapable of it;
but as things are it has some disposition and cause and source of such
an affection, and sometimes it seems to be by having something, but
other times by lacking something that it is such. But if a deprivation
is somehow a condition it has, then in all cases it would be by having
something, though if it is not, this would be so only ambiguously, so
that it is capable by having some condition or by having the lack of it,
if it is possible to "have" a lack.) And in one sense a thing is capable
by virtue of something else's (or its, as something else) not having a
destructive power over it, or source of such a thing, Again, in all these
ways a thing is capable either because of something's just happening
to come about or not come about, or because of its doing so well. For
this sort of potency is present even in things without souls,[14] such as
in instruments, for people say that one lyre is capable of sounding but
another is not, if it doesn't sound good.

And incapacity is a lack of potency and of the sort of source that has
been mentioned, either completely or in what would naturally have
it, or even at the time when it would naturally already have it; for it
is not in the same way that we would say a boy, a man, and a eunuch
were incapable of begetting. Also, for each power there is an opposite
incapacity, both to the one that can only set something in motion, and
to that which can do so well.

And while some things are said to be incapable in this sense of incapacity, others are said in another way to be possible and impossible:
impossible that of which the contrary is necessarily true (in the way

1019b

1019b 10

1019b 20

---

[14] This striking qualification shows that the potencies described so far are found
primarily in living things. They are innate strivings which will emerge so long as nothing
prevents them, described in the *Physics* as inherently yearning for and stretching out
toward form (192a 18). Though the weaker sense of mere possibility is described later
in the chapter, the stronger sense of *dunamis* should always be presumed in Aristotle's
writings.

that it would be impossible for a diagonal to be commensurable be-cause such a thing is false and the contrary of it is not only true but also necessary, so that its commensurability is not only false but nec-essarily false), while, contrary to this, something is said to be possible whenever it is not necessary that its contrary be false, in the way that

1019b 30   it is possible for a human being to sit, since it is not by necessity that his not sitting is false. So in one sense, as has been said, the possible means that which is not necessarily false, in one sense it means what is true, and in one sense it means that which admits of being true. But it is in an altered sense that a power is spoken of in geometry.[15] So these things are possible not as a result of a potency; but other things that

1020a   are spoken of as resulting from a potency are all meant in relation to the one primary sense of the word, that is, a source of change in some-thing else or as something else. For other things are said to be capable either because something has such a power over them, or because it does not have it, or because it has it in such-and-such a way. And it is similar with things that are incapable. Therefore the authoritative definition of the primary kind of potency would be a source of change in something else, or as something else.[16]

Chapter 13   A thing is said to be *so-much* which is divisible into constituents of which each is a one and a *this* by nature. A particular so-much is a multitude if it is countable, but a magnitude if it is

1020a 10   measurable. And a multitude means what is potentially divisible into noncontinuous parts, a magnitude what is divisible into continuous ones; and one sort of magnitude that is continuous in one direction is length, another sort, of what is continuous in two directions, is width, and another sort, of what is continuous in three, is depth. Of these, finite multitude is number, finite length a line, finite width a surface, and finite depth a bodily solid.

Further, some things are said to be so-much in their own right, but other things incidentally, as a line is a certain amount in its own right,

---

[15] And it is by a further alteration and inversion that we speak of the products of the numerical measures of geometrical lines as powers. The Greek word refers to the line itself, as potentially the side of a square or a cube, and thus roughly corresponds to our use of the word root.

[16] This repeated qualification throughout this chapter is meant to distinguish a potency from a nature. See Ch. 4 above.

but an educated thing is so incidentally. And among the ones that are so in their own right, some are so by virtue of their thinghood, in the way that a line is so-much (for in the articulation that says what it is there is present a certain so-much), but others are attributes and states of that 1020a 20 sort of independent thing, such as the many and the few, the long and the short, the wide and the narrow, the deep and the shallow, and other such things. And also the large and the small, and the greater and the less, spoken of both in themselves and in relation to each other, are in their own right attributes of what is so-much; these names, however, are also transferred to other things.[17] But among things that are said too be so-much incidentally, some are so called in the way that the educated and the white were said to be, because that to which they belong is of a certain amount, but others in the way that motion and time are; for these are said to be of certain amounts and continuous 1020a 30 because those things of which these are attributes are divisible. I mean not the thing moved but that through which it was moved, for because that is so-much, the motion too is so-much, and the time because that is.

Chapter 14   *Of what sort* something is said to be means in one way the specific difference of its thinghood, in the sense that a human being is a certain sort of animal because it is two footed, but a horse because it is four footed, and a circle is a certain sort of figure because it is without angles, as though a quality is the specific difference corresponding to 1020b the thinghood. And while this is one way that quality is meant, in another way it is attributed to motionless and mathematical things, as numbers are of certain sorts, such as composite numbers,[18] which are not only along one line but have the plane and the solid as images (and these are so-many-times-so-many or so-many-times-so-many-times-so-many), and in general what is present in the thinghood besides quantity; for the thinghood of each is what it is once, in the way that the thinghood of things that are six is not what they are twice or three times, but what they are once, since six is once six. Also, all things that are attributes of moving things are qualities, such as hotness and

---

[17] An example is a long action (*Categories* 5b 4).

[18] These are all the numbers that are not prime. They sort themselves into many quasi-visible kinds, including not only square numbers, but also triangular ones, pentagonal ones, etc., as well as logical classes such as even-times-even, etc.

1020b 10   coldness, or whiteness and blackness, or heaviness and lightness, and whatever is such, according to which bodies are said to alter when they change. Again, things are said to be of-this-sort in relation to excellence and deficiency or to the good and bad in general.

So of-this-sort could be meant in pretty much two ways, and of these one is the most authoritative; for the primary sense of quality is the specific difference of the thinghood (and quality among number is a part of this, for it is a certain specific difference of independent things, but either not of moving things or of them not as moving), while the other senses are the attributes of moving things as moving, and the specific differences of motions. Excellence and deficiency form

1020b 20   one part among the attributes, since they reveal distinctions of motion and of being-at-work, as a consequence of which the things that are in motion act and are acted upon in a good or an indifferent way; for being able to move and be at work in such-and-such a way is good, whole doing so in such-and-such a contrary way is deficient. But most of all, good and bad signify of what sort something is in the case of things with souls, and of these most of all among those that have choice.

Chapter 15   Some things are said to be *relative* in the way that double is relative to half or triple to one-third, or generally the multiple to what is one of many parts, or what exceeds to what is exceeded; others are meant in the way that what can heat is relative to what can

1020b 30   be heated, or what can cut to what can be cut, and generally what is active to what is passive; others are meant in the way that what is measured is relative to its measure, or what is knowable to knowledge, or what is perceptible to perception.

The first sort are meant in reference to number either simply or determinately, relative either to given numbers or to one. (For example, the double is a determinate number in relation to one, while the multiple is related to one by number but not a determinate number

1021a   such as this or that, but the half-again is related to what it is half-again by a number in relation to a determinate number, while what is a fraction more is related to what it is a fraction more than by an indeterminate number, in just the way that the multiple is to one; but what exceeds is related to what it exceeds in a way that is completely undetermined by number. For number is commensurable, and what is not commensurable is not called a number, but what exceeds in relation to what it exceeds is both that much and more, and this more is indeterminate, since it is whichever of the two it happens to be, either an equal or not an equal part.[19]) So all these are spoken of as relative by reference to number and are attributes of number, as are also the equal and the similar and the same, but in another way. (For they are all meant in reference to a *one,* since those things are the same of which the thinghood is one, similar of which the quality is one, and equal of which the amount is one; but one is the source and measure of number, so that all these things are called relative by reference to number, but not in the same way.)

Things are called active and passive by reference to active or passive potency and to the being-at-work of these potencies; for example, what causes heat is related to what can be heated because it is capable of heating it, and again what is heating is related to what is being heated or what is cutting to what is being cut as being at work upon it. But there is no being-at-work of numerical relations other than in the way that is mentioned in another place,[20] since activities that involve motion do not belong to them. And of relations by way of potency, some are meant as related to what has been acted upon and what will act to what will be acted upon. For it is also in this way that a father is said to be a father of a son, since the one is a thing that has acted, and the other a thing that has been acted upon. Further, some things are related by a lack of potency as being incapable and whatever is said in that way, such as the invisible.

So the things that are said to be relative by number or potency are relative because the very things that they are refer to something else, but not because something else is related to them, but what is measured or known or thought about is said to be relative because something else refers to it. For the thinkable indicates that there is thinking about it, but the thinking is not relative to that of which it is the thinking[21]

---

[19] If the remainder is not commensurable with the whole, no number of equal fractional parts of it can equal the whole. Quantitative relations are therefore of three main kinds: determinate and numerical, numerical but indeterminate, and incapable of any numerical determinacy. Ever since Descartes, in his *Geometry,* collapsed the ideas of multitude and magnitude into one in the "number line," we have lost the last of these distinctions, speaking not of what has no numerical ratio, but of "irrational ratios."

[20] See *Physics* 263a 23–263b 7, where Zeno's famous infinity of half-distances is shown to be present only potentially.

[21] What is thought about is something in itself and not inherently a thought-object.

1021b   (for the same thing would be being said twice), and similarly sight is
             the sight of something, but not of that of which it is the sight (though
             it is true to say this), but relative to color or to some other such thing.
             But in the former way the same thing will have been said twice, that
             it is the sight of that of which it is the sight.

             Now things that are called relative in their own right are in some
             cases so called in these ways, but in others because the classes to which
             they belong are of this sort; for example medical skill is among relative
             things because its genus, knowledge, seems to be something relative.
             Also, things are called relative by reference to the things that have
             them, such as equality because of the equal or similarity because of
             the similar. Other things are relative incidentally, as a human being is
1021b 10   relative because he happens to be double of something, and this is one
             of the relations, or the white is relative if the same thing is incidentally
             double and white.

             Chapter 16   *Complete* means, in one sense, that of which it is
             impossible to find even one of its parts in any way outside it (as the
             complete time of each thing is that outside of which it is not possible
             to find any time which is part of that one), and also means that which
             has nothing of its kind exceeding it in excellence or rightness, as
             someone is called the complete doctor or the complete flutist when
             they lack nothing of the excellence appropriate to their kinds (and by
             transferring this meaning to bad things, we speak of a perfect slanderer
1021b 20   or a complete thief and therefore even call them good, a good thief or
             a good slanderer). And excellence is a certain completeness, for each
             thing is complete and every sort of thinghood is complete at the time
             when the form of its proper excellence lacks no part of the fullness
             it has by nature. And further, those things are said to be complete to
             which a good end belongs, since it is by having the end that they are
             complete, and so, since the end is one of the extremes, transferring
             the meaning, we speak of degenerate things as completely ruined or
             completely decayed, when they lack nothing of ruin and evil but are
             at the extreme point of them. And for this reason even death is by a

             _____

             The relation is therefore not mutual, like the other two kinds, but one sided. See the
             note at the end of Bk. IV, Ch. 5.

transference of meaning called an end, because both are extremes, and
the end for the sake of which something *is* is an extreme.   1021b 30

So things that are called complete in their own right are meant in
that many ways, some by lacking nothing with respect to rightness,
having no superior, and there being nothing to find outside them, but   1022a
others entirely on account of having no superior in their own kinds
nor having anything outside them. The rest result directly from these
either by making something be that way, or having something of that
sort, or being appropriate to some such thing, or in some way or other
being applied to the things that are called complete in the primary
sense.

Chapter 17   *Limit* means the extremity of each thing, the first thing
outside of which there is nothing to find and the first thing inside
of which everything is, which is also the form of a magnitude or of
something that has magnitude; and it means the end of each thing
(and of this sort is that toward which its motion or action tends, but
not that from which it starts, though sometimes it is both and consists
of that from which as well as that toward which), and that for the
sake of which it is, and the thinghood of each thing, and what it is for
each thing to be; for the latter is a boundary of knowledge, and if of   1022a 10
the knowledge, also of the thing. It is clear, therefore, that in however
many ways *source* is meant, limit too is meant, and in yet more ways;
for a source is a limit, but not every limit is a source.

Chapter 18   *That by which* is meant in many ways; in one sense it
is the form or thinghood of each thing, as that by which something is
good is the good itself, but in a sense it means the primary thing in
which something naturally comes to be, such as color in a surface. So
the primary meaning of that by which is the form, but secondarily it
is used of the material of each thing and the primary thing underlying
it. But in general *that by which* will be appropriate in the same number
of ways as *cause*; for one says either "by what reason did he come?" or   1022a 20
"for the sake of what did he come?" and either "by means of what did
he make a mistake or draw a conclusion?" or "what is the cause of the
conclusion or of the mistake?" And also by which is used by reference

to a position which one stands by or walks by,[22] since all these uses indicate a position or a place.

Therefore *by itself* must also be meant in many ways. For in one sense the by itself is what it is for each thing to be; for instance Callias by virtue of himself is Callias and what it is for Callias to be. But in a sense it means all the things that are present in what something is, as Callias by virtue of himself is an animal, since animal is a constituent in his articulation because Callias is a certain kind of animal. And also it means anything that something receives in itself primarily, or in any of its parts, in the way that a surface is white by virtue of itself or a human being lives by virtue of himself; for the soul is a part of the human being, and in it living primarily resides. Also it means that for which no other thing is responsible; for there are many causes of a human being, such as animalness and two-footedness, but still by himself a human being is a human being. Also it means all those things that belong to something alone, insofar as it is alone, for which reason what is separated is by itself.

Chapter 19   *Disposition* means an ordering of something that has parts, either in place or in power or in kind; for it has to have some sort of position, as the name disposition indicates.

Chapter 20   An *active state* of something is meant in one sense as a certain being-at-work of the thing that has it and what it has, just as if it were a certain action or motion (for whenever one thing does something and another has it done to it there is a doing shared between them, and so too between someone who has clothes on and the clothes he has on there is a shared state of having). So in this sense it is clear that it is not possible to *have* a state of having (since it would go on to infinity if there were to be a having of the state of having what something has). But in another sense an active state means a disposition by which the thing disposed is in a good or bad condition, either in its own right or in relation to something else, in the way that health is an active state, since it is that sort of disposition. And it is also called an active state if

1022a 30 / 1022b / 1022b 10

---

22 More properly "at which one stands or along which one walks." The range of meanings of *kata* has no single English equivalent, but the translation chosen here is the simplest one that most nearly fits the most uses.

there is a part of such a disposition, for which reason the excellence of the parts is also a certain active state.

Chapter 21   An *attribute* means in one sense a quality in respect to which something is capable of being altered, such as white and black, or sweet and bitter, or heaviness and lightness, or whatever else is of this sort. But in a sense it is used of these when they are at work and have already been altered. But more than these, it implies harmful alterations and motions, and especially painful harm.[23] Also, great misfortunes are called suffering.

1022b 20

Chapter 22   *Lacking* is meant in one way of what does not have something that it is natural to have, even though it is not *it* that is of such a nature as to have it, in the way that a plant is said to lack eyes; and in a sense it is used of what does not have something that is natural to have for either it or the class of things it belongs to, as in different senses a human being who is blind and a mole lack eyes, the one in respect to its class and the other in its own right. Also it is used of what does not have something that is natural to have when it is natural to have it, for blindness is a lack, but one is not blind at any age but only if one does not have sight at the age when it is natural to have it. Likewise, it is used of what does not have something in the conditions, or by the means, or in the relation, or in the manner that it is natural to have it. And the violent taking away of anything is called a depriving.

1022b 30

And in as many ways as negations are expressed by means of prefixes and suffixes, in so many ways are deprivations also expressed. For a thing is called unequal for not having an equality natural to it, or invisible for either not having any color at all or having a faint one, or footless for either not having feet at all or having inadequate ones. Also, negations are used for what has little of something, such as seedless fruit, and this is in a way to have it inadequately. And they are also used of what has something but not easily or well, as a thing is called uncuttable not only for not being able to be cut, but also for

1023a

---

[23] Chapters 19–21 form a cluster. A disposition may be transitory but an active state holds on and continues; an attribute may also be enduring but is a passive state. But the meaning of *pathos* spans all passive conditions from the most indifferent to great afflictions; no English word hits both notes in a similar way.

not being able to be cut easily or well. Still, they sometimes mean what does not have something in any way, for someone who is one eyed is not called blind, but only one who lacks sight in both eyes, for which reason not everyone is good or bad, or just or unjust, but there are also in-betweens.

Chapter 23   *Having* is meant in many ways, in one sense as keeping something else in accordance with something's own nature or impulse, for which reason a fever is said to have hold of a human being, or tyrants to have hold of cities, or those wearing clothes to have them on. But in a sense it means that in which something is present as a thing receptive of it, as bronze has the form of a statue, or the body has a disease. And in a sense it is used in the way that what surrounds has what is surrounded, for a thing is said to be had by that in which it is contained, as we say a pitcher has a liquid, a city has people, or a ship has sailors, and in this sense too, the whole has the parts. Also, what obstructs something from moving or acting by its own impulse is said to have hold of it, as columns have hold of the heavy things that press down on them, and as poets make Atlas hold heaven, as though it would fall down to earth, just as some of the writers about nature also say. And in this way too what is continuous is said to hold together what it connects, as though it would be separated apart by each part's own impulse. And being-in something is meant in ways that reflect and correspond to having.

Chapter 24   To be *from* something means in one sense that a thing is made out of it as material, and this in two ways, as either the highest genus or lowest species, since there is one sense in which all things that melt are made of water, but another sense in which a statue is made of bronze. But in a sense it is meant in the way that something comes from a first source that sets it in motion. (For example, from what did the fight come? From insults, because these were the source of the fight.) And in a sense it means the way that something is from the composite of material and form, just as the parts are from the whole and the verse is from the *Iliad* and the stones are from the house; for the form is an end, and it is what has its end that is complete. And in some cases it is meant in the way that the form is derived from its part, as humanness from two-footedness and the syllable from the letter; for this is so in different way from that in which the statue is made

1023a 10

1023a 20

1023a 30

1023b

of bronze, since the composite independent thing is made of sensible
material, but the form is made of intelligible material. So some things
are meant in these ways, but others because something is applicable
to these ways by means of some part, as the child comes from a father
and mother or plants from the earth because they come from some part
of them. And in a sense it means that after which something comes in
time, as night from day or a storm from a calm, because the one is after
the other; and among these, some are spoken of in this way for having
a change into each other, as in the things just mentioned, but others
for being successive in time only, as the sailing was from the equinox   1023b 10
because it started after the equinox, and the Thargelian festival comes
from the Dionysian because it is after the Dionysian.

Chapter 25   A *part* means in one sense that into which a quantity
is divided in any way whatever (for always what is taken away from a
quantity as a quantity is said to be a part of it, as two is in a way called
part of three), but in another sense it means only those among such
things that measure off the whole without remainder, so while two is
in one way called part of three, in another way it is not. Also, the things
into which the form is divided apart from quantity are called parts of
it, for which reason people say that the species are parts of the genus.
Also, it means the things into which a whole is divided or of which it is
composed, the whole being either a form or the thing that has a form;   1023b 20
for example, of a bronze sphere or a bronze cube, both the bronze (that
is, the material in which the form is present) and the angle are parts.
Also, it means the things that are in the articulation that reveals each
thing, and these are parts of the whole, for which reason the genus is
also called part of the species, though in another way the species is
part of the genus.

Chapter 26   A *whole* means that of which no part is absent out of
those of which it is said to be a whole by nature, or that which includes
what it includes in such a way that they are some one thing. The latter
can happen in two ways: either such that each is one or so that the
one thing is made out of them. For a universal, which is attributed to
a whole class of things as though it were a certain whole, is universal   1023b 30
in the sense that it includes many things by being predicated of each,
and that they are all one each-by-each, as are a human being, a horse,
and a god, because they are all living things; but what is continuous

and finite is a whole whenever some one thing is made of a number of things, most of all when they are distinct constituents of it only potentially, but if not, actively as well. And among these themselves, those that are of this sort by nature are wholes more so than are those that are so by art, just as we also said in the case of oneness, since wholeness is a certain kind of oneness. Also, what has quantity has a beginning, middle, and end, and any quantity in which position makes no difference is spoken of as "all," while any in which it does make a difference is called a whole. All those that admit of being both ways are both whole and all, and these are all those things of which the nature stays the same through a rearrangement but the form does not, such as wax or a cloak, for each of these is spoken of both as a whole and as all, since it has both conditions. But water, and as many things as are liquid, and a number are spoken of as all, but one does not speak of the whole number or the whole water except by some altered meaning. And everything that is spoken of as all when it is referred to as one thing, is spoken of as "every" when it is referred to as divided up: all this is a number, but every one of these is a unit.

1024a

1024a 10

Chapter 27 *Defective* is applied to things that have quantity, but not to any random one, since it has to be made of parts and also be a whole. For two is not defective if one of its ones is taken away (for the defective thing is never equal to the part left off), nor is any number at all, for it is also necessary that an independent thing persist; for a cup to be defective it must still be a cup, but a number is no longer the same thing. And on top of these things, even if something is of unlike parts, not even all of these can be defective (since a number too in a way has unlike parts, such as two and three), but in general, none of those things in which position makes no difference can be defective, such as water or fire, but such things have to be of the sort that have position on account of their thinghood. Also, they must be continuous, for a harmony consists of unlike parts that have position, but it cannot become defective. And beyond this, not even those things that are wholes are defective by lacking any part whatever. For it is necessary that they lack neither the parts that are decisive to their thinghood nor those that are just of any sort whatever; for example, if a cup is pierced it is not defective, but it is if it lacks its handle or any prominent part, and a human being is not defective if he lacks some flesh or his spleen, but he is if he lacks some extremity, and not just any of those

1024a 20

but one which cannot grow back when it is completely removed. For this reason bald men are not defective.

Chapter 28   A *kind* is spoken of in one way if there is continuous generation of things that have the same form, in the sense that "as long as there is a human kind" means "as long as there is a continuous generation of human beings" and in another way it means the first mover from which things were brought into being, for it is in this way that Hellenes or Ionians are spoken of as a kind, on account of being descended from either Hellen or Ion as a first progenitor, and this is more the case with those who are descended from a progenitor than those from a material stock (though a kind is also spoken of as coming from the female line, as with the descendants of Pyrrha). And it is meant in the way that plane is the kind to which plane figures belong and solid is that of solid ones, for each figure is either such-and-such a plane or such-and-such a solid, and this is what underlies their specific differences. It also means the primary constituent in the articulation of something, which is stated in saying what it is; for this is its kind, of which its qualities are stated as specific differences. So kind is meant in that many ways: by reference to continuous generation of the same form, to a first mover that is alike in form, or as material. For that to which the specific difference and the quality belong is some underlying thing, which we call material.

And different in kind means things which have different things primarily underlying them, which cannot be reduced one to the other or both to the same thing; for example, form and material are different in kind, as are all those things that are meant in reference to different ways of predicating being (for some things signify what beings are, others of what sort they are, and others the things that were distinguished above[24]), since these are not reducible at all, neither into one another nor into any one thing.

Chapter 29   *False* is meant in one way in the sense that a thing is false, and in this way one sort of falsity is because the thing doesn't go together or cannot be put together (in the way that one claims the diagonal to be commensurable or you to be sitting down, for one of

1024a 30

1024b

1024b 10

1024b 20

---

[24] At 1017a 23–27.

these is false always and the other sometimes, since these things are not that way), and another sort includes things which are, but which naturally seem to be either of a sort that they are not, or things that are not (for instance a perspective drawing or dreams, for these are something, but not what they produce the appearance of). Things, then, are said to be false in this way, either because they are not, or because the appearance of them is of something that is not.

Now a statement is false which is about things that are not, insofar as it is false, for which reason every statement is false about something other than that about which it is true, as what is true of the circle is false about the triangle. And of each thing there is a sense in which there is one statement, which says what it is for it to be, but there 1024b 30 is another sense in which there are many statements, since the thing itself and the thing when it has been affected are in a certain way the same, such as Socrates and the educated Socrates (but a false statement is not a statement about anything simply). For this reason Antisthenes believed simplemindedly that nothing was fit to be said except a thing's own articulation, one statement for one thing, from which it followed that there could be no contradicting, and practically no making a mistake. But is it possible to speak of each thing not only by its articulation but also by that of something else, and this either 1025a completely falsely, or it is possible in a way also truly, as eight is called double by means of the articulation of two.

Some things, then, are called false in these ways, and a false human being is one who skillfully and deliberately makes use of such statements, for no other reason but for its own sake, and who foists such statements onto other people, in the same way that we also said that things are false which produce a false appearance. For this reason the argument in the *Hippias*[25] goes wrong in saying that the same person is false and true. For it takes the one who is capable of deceiving as being false (and this is the one who knows and has understanding), and further assumes that the one who willingly does low things is better 1025a 10 than one who does so unwillingly. But this takes something false out of a survey of examples, for the one who limps willingly is better off

---

[25] Plato's *Lesser Hippias*. Aristotle is not arguing with Plato, but answering Socrates's invitation to see through the confusion in Hippias's thinking. Hippias is depicted as an empty-headed professor who can speak expertly about Homer but cannot make elementary distinctions.

than the one who does so unwillingly if limping means imitating a limp, since if one is in fact willingly lame he is perhaps worse off, just as in the case of moral character, in this one too.

Chapter 30   *Incidental* means what belongs to something and is true to say of it, but is not so either necessarily or for the most part, such as if someone who dug a hole for a plant found a treasure. This finding a treasure is incidental to the one who dug the hole, for it is neither by necessity that the one comes from the other or with the other, nor for the most part that one who plants something finds a treasure. And someone with a refined education might be of a pale complexion, but since this happens neither by necessity nor for the most part, we call it incidental. Therefore, since there is something that belongs and something it belongs to, and some of these are present at a certain place and time, whatever belongs to something, but not because it is this thing or at this time or in this place, will be incidental. And so there is not any definite cause of what is incidental, but a chance cause, and this is indeterminate. It happened incidentally to someone to go to Aegina, if he arrived not because of an intention to go there, but because he was driven off course by a storm or taken by pirates. The incidental thing is indeed something that has happened or *is*, but by virtue not of itself but of something else; for the storm was the cause of his going somewhere he was not headed for.

1025a 20

1025a 30

But incidental is also meant in another way, as of all those things that do belong to each thing in virtue of itself but are not present in the thinghood of them; for instance, it is incidental to a triangle to contain two right angles. Incidental attributes of this kind can be everlasting, but none of the former kind can be. But an account of this is in other writings.[26]

---

[26] This second type of incidental attribute is proper (*idion*) to the thing it belongs to, and in the logical writings it is called by that name and usually translated as "property." See for example *Topics* 102a 18–31. In the strictest sense, what a triangle is in its own right, in virtue of itself, is just a plane figure with three angles and straight sides. Because it is that, it also incidentally contains two right angles, and even though that property is both necessary and sufficient to identify a triangle, it is derivative from what makes it a triangle.

1025b 30   be and its articulation in speech not be ignored, since inquiring without this is doing nothing at all. But among things that are defined and the kinds of what-it-is of things, some are like the snub and other like the concave. And these differ because the snub is conceived along with its material (since what is snub is a concave nose), while concavity is without sensible material. So if all natural things are meant in a way

1026a   similar to the snub, as for example nose, eye, face, flesh, bone, animal in general, leaf, root, bark, and plant in general (for the meaning of none of them leaves out motion, but they all have material), it is clear how one must look for and define what is it for natural things to be, and also why it belongs to the one who studies them to pay attention to some aspect of the soul, namely as much of it as is not without material.²

That, then, the study of nature is contemplative is clear from these things, but mathematics is also contemplative, though whether it is about things that are motionless and separate is not now clear; it is clear, however, that it studies some mathematical things insofar as they

1026a 10   are motionless and insofar as they are separate. But if there is anything that is everlasting and motionless and separate, it is obvious that the knowledge of it belongs to a contemplative study, though surely not to the study of nature, nor to mathematics, but to one that precedes them both. For the study of nature concerns things that are indeed separate, but are not motionless, while some mathematics concerns things that are indeed motionless, but presumably not separate, but in truth in material; but the first contemplative study concerns things that are both separate and motionless.

And while it is necessary that all causes be everlasting,³ these are so most of all, since they are responsible for what appears to us of the divine. Therefore there would be three sorts of contemplative philosophy, the mathematical, the natural, and the theological; for

1026a 20   it is not hard to see that if the divine is present anywhere, it is present in a nature of this kind, and that the most honorable study must be

---

² A human being is not simply natural, for while soul and body are inseparably one, just like wax and the shape impressed in it (*On the Soul* 412b 6–8), the intellect has no bodily organ (429a 26–27), but comes in "through the door" (*Generation of Animals* 736b 27–28).

³ See 1013a 17–23. Since all causes are sources, they cannot have other sources, but must be self-explanatory and self-sustaining.

about the most honorable class of things. The contemplative studies, then, are more worthy of choice than are the other kinds of knowledge, and this one is more worthy of choice than are the other contemplative studies.

One might raise an impasse whether first philosophy is universal or is concerned with a certain class of things and some one nature (for these are not the same way of going about things in mathematics either, but geometry or astronomy is about a certain kind of nature, while universal mathematics is common to them all[4]). Now if there were no other independent thing besides the composite natural ones, the study of nature would be the primary kind of knowledge; but if there is some motionless independent thing, the knowledge of this precedes it and is first philosophy, and it is universal *in just that way,* because it is first.[5] And it belongs to this sort of philosophy to study being as being, both what it is and what belongs to it just by virtue of being.

1026a 30

Chapter 2   But since being, spoken of simply, is meant in more than one way, of which one is incidental, another is as the true (and nonbeing as the false), and besides these there are the modes of predication (such as what, of what sort, how much, where, and when something is, and anything else "is" means in this way), and still besides all these being-potentially and being-at-work—since, that is, being is meant in so many ways, one ought first to say about the incidental kind that there is no contemplative study of it. For it is a sign of this that it is not the business of any kind of knowledge to be concerned with it, neither the knowledge that aims at action, nor that which aims at making something, nor that which aims at contemplation. For the one who makes a house does not make all the things that come along incidentally with the house at the same time that it comes into being (for they are infinite, for nothing prevents what is made from being pleasant to some people, obnoxious to others, and useful to still others, and different, one might say, from everything that is, with none of which the housebuilding craft is concerned), and in the same way, the

1026b

1026b 10

---

4 This sort of mathematics may be found in Book V of Euclid's *Elements.*

5 See Bk. IV, Ch. 1, and its footnote. Being as being is not the lowest common denominator of all that is, universal because nothing lacks it, but the highest, divine, kind of being, universal because everything else depends on it for its being.

geometer does not study the things that are incidental in this way to figures, nor whether "triangle" is different from "triangle containing two right angles." And this is in accord with good sense, for the incidental is as though it were only a word.

For this reason Plato's assignment of sophistry to what concerns nonbeing was in a certain way not bad.[6] For the arguments of the sophists are, one might say, most of all about what is incidental: whether educated and literate are different or the same, and whether educated Coriscus is the same as Coriscus, and, if everything that is, but is not always, has come into being, that it follows that, if one who was educated became literate, then also one who was literate became educated, and all the other arguments that are of this kind. For it is obvious that what is incidental is something close to what is not. And this is clear also from this sort of argument: there is coming into being and passing away of things that are in some other sense, but not of things that are incidentally.[7] But one ought, nevertheless, still to say about the incidental, to the extent that it is possible, what the nature of it is and through what cause it is, for at the same time it will presumably also be clear why there is no knowledge of it.

1026b 30

Since, then, among beings, some are always and necessarily in the same condition, meaning this not in the sense that it is by force but as what is not capable of being otherwise, while others are as they are neither necessarily nor always, but for the most part, this is the source and this is the cause of there being what is incidental; for what is neither always nor for the most part is what we say is incidental.[8] For instance, if a storm and cold weather should come about in the dog days of summer, we say that this is incidental, but not if there is heat and stifling air, since the latter happens always or for the most part, but the former does not. And it is incidental to a human being

---

[6] This is the seventh and final definition of the sophist in the dialogue of that name, which depends upon understanding an image as a way in which nonbeing is present. More than half the dialogue, from 263E on, is devoted to justifying this claim.

[7] One of Aristotle's examples elsewhere is of a housebuilder who plays the flute. Your house may happen to have been built by a flute player, but it didn't come into being as such, since the builder's knowledge of flute playing didn't direct the making of it. Incidental attributes just attach to things, without having been worked into the texture of them.

[8] That is, since the regularity of the things around us is not always rigid, there is room around the edges of things for all sorts of incidental attributes.

to be pale (since it is so neither always nor for the most part), but it is not incidentally that he is an animal. And for a housebuilder to cure someone is incidental, because it is not natural for a housebuilder but for a doctor to do that, but the housebuilder incidentally was a doctor. And a fancy chef, aiming at pleasure, might make something healthful, but not on account of his skill as a fancy chef; for this reason it was incidental, we say, and there is a sense in which he made the healthful food, but in the simple sense of "make" he did not.

For some things are results of capacities to produce other things, while others result from no definite art or capacity; for of what is or happens incidentally, the cause too is incidental. Therefore, since not all things are or happen necessarily and always, but most things are and happen for the most part, it is necessary that there be incidental being. For instance, it is neither always nor for the most part that someone pale has a refined education, but since it sometimes happens, it will be incidental (or if not, everything would be by necessity). Therefore, it will be the material that is capable of being otherwise than it is for the most part, that is the cause of what is incidental. But one must take this as a starting point, whether there is nothing that is neither always nor for the most part the case, or whether this is impossible. So there is something besides these that is whichever way it chances to be and is incidental.

But while there are things that are so for the most part, does nothing belong to anything always, or are some things everlasting? One must examine these things later, but it is clear that there is no knowledge of what is incidental, since all knowledge is of what is so always or for the most part—for how else will anyone learn or teach? For it is necessary to make something definite by means of what it is always or for the most part, such as that milk-and-honey is for the most part beneficial to one who has a fever—but what happens contrary to this one will not be able to say, when it is not so, such as on the new moon; for what happens on the new moon will also be the case either always or for the most part, but the incidental is something besides these. What, then, the incidental is, and through what cause it is,[9] and that there is no knowledge of it, have been said.

---

[9] That the material of things is capable of being other than it usually is makes room for incidental attributes, and so, while it is not a sufficient cause for their occurrence, it secures their possibility.

1027a 30   **Chapter 3**   That there are sources and causes which come and go without being in a process of coming-into-being or passing-away is evident. For if this were not so, all things would be by necessity, if there must be some nonincidental cause of what is coming into being or passing away. For will this particular thing be the case or not? It will if this other thing has happened, but if not, not. But that one will happen if some other thing has happened. And it is clear in this way

1027b   that by always taking some time away from a finite time one will come to the present, so that this person will die by violence, if he goes outside, and will do this if he gets thirsty, and that if something else happens, and in this way one will come to what is now present, or to something that has happened. For example, he will get thirsty if he eats spicy food, and this is either the case or not, so that it is by necessity that he is killed or is not killed. And similarly if one jumps back to the past, the account is the same; for this already belongs to something (I mean the thing that has happened). Therefore everything that is going to happen will be by necessity. For example, one who is alive

1027b 10   will necessarily die because of something that has already come about, such as contrary tendencies in the same body. But whether he will die by disease or violence is not yet necessary, but only if some particular thing has happened. Therefore it is clear that the result goes back as far as some starting point, but this no longer goes back to anything else. This, then, will be the origin of what happens in whichever way it chances to, and nothing else will be responsible for its happening. But to what sort of source and what sort of cause such tracing back has gone, whether to material or to that for the sake of which or to a mover, one needs to examine with the greatest care.[10]

Chapter 4   Let the things that concern incidental being be set aside,

---

[10] The preceding chapter argued that a material cause of anything incidental must be present, and this chapter shows how the other two kinds mentioned here might be more governing. The spicy food is an external source of motion, but the human being still chooses to go outside for the sake of getting water, and this final cause is primary in that example. In his most thorough account of chance, *Physics* Bk. II, Ch. 4–6 and 8, Aristotle shows that all chance events, and hence all incidental being, trace back to an interference of final causes. This requires establishing that all natural events are for the sake of ends. Here it is sufficient for his argument to show that incidental being lacks the knowable determinacy of what happens always or for the most part. A brief excerpt from the *Physics,* at the end of Bk. XI, Ch. 8, below, makes explicit that incidental being is always derivative from things that are for the sake of something.

since it has been marked off sufficiently. But since being as the true and nonbeing as the false concern combining and separating, and the whole topic concerns the division of a pair of contradictories (for truth has the affirmation in the case of a combination and the denial in the case of a separation, while the false has the contradictory of this division, but how it happens that one thinks things together or apart is another story—I mean together and apart in such a way that they are not in sequence but become some one thing), since the false and the true are not in things, as if, say, the good were true and the bad automatically false, but in thinking, and not even in thinking in what concerns simple things and what things are, then whatever one needs to pay attention to about this sort of being and nonbeing should be examined later[11]; but since the intertwining and dividing are in thinking but not in things, and being of this sort is different from the being of what *is* in the governing sense (for thinking attaches or separates what something is, or that it is of this sort, or that it is this much, or anything else it might be), both being as what is incidental and being as what is true must be set aside. For the cause of the one is indeterminate and of the other is some attribute of thinking, and both kinds concern the remaining kind of being, and do not reveal any nature that is outside this sort of being; for this reason let them be set aside, but what must be examined are the causes and sources of being itself, as being.

1027b 20

1027b 30

1028a

---

[11] See Bk. IX, Ch. 10.

Book VII (Book Z)
Thinghood and Form[1]

Chapter 1   Being is meant in more than one way, just as we distinguished earlier in the chapters concerning the number of ways things are meant.[2] For it signifies what something is and a *this*, but also of what sort or how much something is, or any of the other things attributed in that way. But while being is meant in so many ways, it is obvious that the way that is first among these is what something is, which indicates its thinghood (for whenever we say that this or that is of a particular sort, we say that it is either good or bad, but not three feet long or a human being, but when we say what it is, we say not that it is white or hot or three feet long, but a human being or a god), and the other kinds of being are attributed to something that *is* in this way, some of them as amounts of it, others as qualities of it, others as things that happen to it, and others as something else of that kind. And for this reason, someone might be at an impasse whether each thing such as walking or healing or sitting is or is not a being, and similarly with anything else whatever of such a kind; for none of them is either of such a nature as to be by itself nor capable of being separated from an independent thing, but instead, if anything, it is the thing that walks or sits or gets well that is one of the beings. And it is obvious that these *are* more so, because there is something determinate that underlies them (and this is the independent thing or the particular), which is reflected in a predicate of such a kind; for the good or the seated are not meant apart from this. So it is clear that each of those former things *is* by means of this one, so that what is primarily, not being in some particular way but simply being, would be thinghood.

But then primary is meant in more than one way, but all the same, thinghood is primary in every sense, in articulation, in knowledge, and in time. For none of the other ways of attributing being is separate, but only this one; and in articulation this one is primary (for in the articulation of anything, that of its thinghood must be included); and we believe that we know each thing most of all when we know *what* it

---

[1] This title for Book VII supplied by the translator.

[2] See Bk. V, Ch. 7.

1028b

is—a human being or fire—rather than of what sort or how much or where it is, since we know even each of these things themselves only when we know *what* an amount or a sort is. And in fact, the thing that has been sought both anciently and now, and always, and is always a source of impasses, "what is being?", is just this: what is thinghood? (For it is this that some people say is one and others more than one, and some say is finite and others infinite.) So too for us, most of all and first of all and, one might almost say, solely, it is necessary to study what this kind of being is.

Chapter 2   Now thinghood seems to belong most evidently to
1028b 10   bodies (and therefore we say that animals and plants and their parts are independent things, as well as natural bodies such as fire and water and earth and each thing of that kind, and as many things as are either parts of these or made out of them, out of either some or all of them, such as the cosmos and the parts of it, the stars and the moon and the sun). But whether these alone are independent things or there are also others, or just some of these are, or some in addition to some other things, or none of these but something different, must be examined. And it seems to some that the limits of bodies, such as a surface and a line and a point and the unit, are independent things more so than are a body or a solid. Further, while some believe that there is no such thing apart from what is perceptible, others believe that there are everlasting things that are more in number and that
1028b 20   *are* more, just as Plato believed that the forms and the mathematical things are two kinds of independent things, while the thinghood of perceptible bodies is a third, and Speusippus believed in still more kinds of thinghood originating from the one and from each source of thinghood, one source for numbers and another for magnitudes, and next one for soul, and in this way he extended the kinds of thinghood. But some people say that the forms and the numbers have the same nature, and that the other things follow upon them, lines and planes all the way down to the thinghood of the cosmos and the perceptible things.

Now about these things, what is said well and what not, and what the independent things are, and whether there are any apart from the perceptible things or not, and in what way these are, and whether
1028b 30   there is any separate independent thing, and why and in what way, or

none at all apart from perceptible things, must be examined by those beginning to sketch out what thinghood is.

Chapter 3 Now thinghood is meant, if not in more ways, certainly in four ways most of all; for the thinghood of each thing seems to be what it keeps on being in order to be at all, but also seems to be the universal, and the general class, and, fourth, what underlies these. And what underlies the others is that to which they are attributed, while it is itself not attributed any further to anything else; therefore one ought to distinguish this sort first, since thinghood seems most of all to be the first underlying thing. And in a certain way the material is said to be of this sort, but in another way the form is, and in a third that which is made out of these. (And by the material, I mean, for instance, bronze, by the form, the shape of its look, and by what is made out of these, the statue.) So if the form is more primary than the material, and *is* more, it will also, for the same reason, be more primary than what is made of both.

1029a

So now, in a sketch, what thinghood is has been said, that it is what is not in an underlying thing but is that in which everything else is; but it is necessary not to say only this, since it is not sufficient, for this itself is unclear, and what's more, the material becomes thinghood. For if thinghood is not this, what else it is eludes us, since, when everything else is stripped away, it does not seem that anything is left; for some of the other things are attributes of bodies, or things done by them, or capacities of them, while length, breadth, and depth are certain quantities but not independent things (for how much something is is not thinghood), but it is rather the first thing in which these are present that is an independent thing. But when length, breadth, and depth are taken away, we see nothing left behind, unless it is what is bounded by these, so that, to those who look at it in this way, the material must seem to be the only independent thing. By material I mean that which, in its own right, is not said to be either something or so much or anything else by which being is made definite. For there is something to which each of these is attributed, and of which the being is different from each of the things attributed (for everything else is attributed to thinghood, and it is attributed to the material), so that the last thing is in itself neither something nor so much, nor is it anything else; and it is not even the negations of these, for these too would belong to it as attributes.

1029a 10

1029a 20

So for those who examine it from these starting points, thinghood turns out to be material. But this is impossible, for also to be separate and a *this* seem to belong to an independent thing most of all, on account of which the form and what is made out of both would seem to

1029a 30   be thinghood more than would the material. And surely the thinghood that consists of both, I mean of the material and the form, should be put aside, since it is derivative and obvious; and in a certain way the material too is evident, but one must investigate about the third kind, since this is the greatest stumbling block. And it is agreed that there are some independent things among perceptible things, so one ought

1029b 3   to look first into these. For it is convenient to pass over toward what is most knowable. For learning happens in this way in all areas, by way of what is less knowable by nature, toward what is more knowable. And this is the task: just as, where actions are concerned, one's job is to make what is completely good be good for each person *out of* the things that are good for each one, so too it is to make what is knowable by nature known to oneself out of the things that are more known to one. But the things that are known and primary to each person are often

1029b 10   scarcely knowable, and have little or nothing of being; nevertheless one must try to come to know the things that are completely knowable *out of* the things that are poorly known but known to oneself, passing over, as was said, by means of these very things.[3]

1029b 1   **Chapter 4**   But since at the start we distinguished in how many ways we define thinghood, and of these a certain one seemed to be what something keeps on being in order to be, one ought to examine

1029b 13   that. And first let us say some things about it from the standpoint of logic,[4] because what it is for each thing to be is what is *said* of it in its own right. For being you is not being cultivated, since it is not in virtue

---

[3] The last three sentences are misplaced in the manuscripts, after the first sentence of Ch. 4. They are undoubtedly genuine, and form one of the major structural connections of the whole *Metaphysics*. They make it clear that the sort of being found in perceptible things, while it must be first for us, is not first in the nature of things. On this order of study, see also *Physics* Bk. I, Ch. 1, and *Topics* 101a 35–101b 4.

[4] The rest of Bk. VII, except for Ch. 7–9, is logical in character, an analysis starting from the way we speak and think. For Aristotle, this is always secondary to examining the way things are by nature. (See, for example, *Physics* 204b 3–11.) The last chapter of Bk. VII forms a bridge from the logical to the natural by means of the notion of cause.

of yourself that you are cultivated. Therefore, being you is what you are in virtue of yourself, but it is not even all of this, for it is not what is in virtue of itself in the way that white is in a surface, because being white is not being a surface. But surely neither is the thing made out of both, being-a-white-surface, what it is to be white, because white itself is attached to it. Therefore that articulation in which something is not itself present, when one is articulating *it,* is the statement of what it is for each thing to be; so if being a white surface is being a surface that is smooth,[5] being, for white and for smooth, is one and the same.

1029b 20

But since there are also compounds that result from the other ways of attributing being (since there is something underlying each of them, such as the of-what-sort, the how-much, the when, the where, and the motion), one must consider whether there is for each of these a statement of what it is for it to be, and whether what-it-is-for-it-to-be even belongs to them, for example to a person with a pale complexion. Now let's suppose the name for a pale person is "sheet." What is the being of a "sheet"? Now surely this is not even among the things attributed to anything in virtue of itself. But "not in virtue of itself" is meant in two ways, and of these one results from attaching something, but the other from not attaching something. For the former way is stated by sticking the thing one is defining onto something else, as if, when defining being-pale, one were to state the articulation of a pale person; the latter way occurs because something else *is* attached to the thing being defined, as if "sheet" meant a pale person and one defined the "sheet" as pale.[6] The pale person is of course pale, but what it is for it to be is not what it is for pale to be.

1029b 30

1030a

But is being-a-"sheet" any sort of what-it-is-for-something-to-be at all, or not? For what it is for something to be is the very thing that

---

[5] See *On Sense Perception and Perceptible Things* 442b 10–13. Democritus, like Galileo and Descartes in a later time, sought to reduce the proper objects of the senses to mathematical attributes. Aristotle does not think white can be reduced to anything else, but his point is that, even though white by its very nature must be in a surface, being-in-a-surface is no part of the nature of whiteness, whatever that nature might be.

[6] The statement of what something is in virtue of itself can fail by including too much, or by omitting something necessary. Aristotle is implicitly asking, in the two preceding sentences, why a "sheet" can't be what it is in virtue of itself, so long as one states that properly, and his answer is in the next paragraph.

something is, and whenever one thing is attributed to another, the compound is not the very thing that is a *this*, as in this instance a pale person is not the very thing that is a *this*, if indeed *this*ness belongs only to independent things. Therefore there is a what-it-is-for-it-to-be of all those things of which the articulation is a definition. And it is not the case that there is a definition whenever a name means the same thing as a statement (for then all statements in words would be definitions, since there could be a name for any group of words

1030a 10   whatever, and even the *Iliad* would be a definition), but only if the statement articulates some primary thing, and things of this kind are all those that are not articulated by attributing one thing to another. Therefore there will be no what-it-is-for-it-to-be belonging to anything that is not a species of a genus, but only to these (for the species seems not to be meant as something a thing has a share in and is affected by, nor as an incidental attribute); but there will still be a statement for each of the other things as well, of what it means, if it has a name, stating that this belongs to that, or a more accurate statement instead of a simple one, but there will be no definition nor any what-it-is-for-it-to-be.

Or else, are the definition and the what-it-is-for-something-to-be both alike meant in more than one way? For also what-something-is in one sense indicates its thinghood and a *this*, but in another sense

1030a 20   indicates any of the ways of attributing being, how much something is, of what sort it is, and everything else of that kind. For just as the "is" belongs to all of them, though not in the same way, but to one of them primarily and to the rest derivatively, so too the what-it-is belongs simply to the thinghood but in a certain respect to the others; for we might also ask what an of-this-sort is, though not simply but in the same way that, in the case of what is not, some people say logically that what is not *is*, not that it is simply but that it is what is not. It is the same way also with what is of-this-sort. Now one ought to consider how one should speak about each thing, but surely not more than about how the things are; so also now, since it is clear what is meant, the what-it-is-for-something-to-be, in the same way as the what-something-is, will also belong primarily and simply to

1030a 30   thinghood, and secondarily to the other ways of attributing being, not as a what-it-is-for-something-to-be simply, but as what it is for an of-this-sort or a so-much to be.

For one has to say that these are beings either ambiguously or by adding and subtracting,[7] in the same way that one can say the unknown is something known, since the right thing to say is that they are called beings neither ambiguously nor in just the same way, just as what is medical is so called not by being something that is one and the same but by pointing to something that is one and the same, and so not in an ambiguous way either. For a medical cadaver, a medical action, and a medical instrument are meant neither ambiguously nor as one thing, but as pointing to one thing.[8] But it makes no difference in which of the two ways one wants to speak about these things. This is clear: that a definition and a what-it-is-for-something-to-be belong primarily and simply to independent things. It is not that they do not belong to the other things in a way that resembles this, but only that they do not belong to them primarily. For it is not necessary, if we assume this, that there be a definition of whatever means the same thing as a group of words, but only that there be one of what means the same thing as a certain kind of group of words, and this is one that belongs to something that is one, not by being continuous in the way that the *Iliad* is, nor by being bundled together, but in just those ways that one is meant. Now one is meant in the same ways as being, and being signifies in one way a *this*, in another a so-much, and in another an of-this-sort. And for this reason there will be a statement and a definition of a pale person, but in a different way than of pale, or of an independent thing.

1030b

1030b 10

Chapter 5   But there is an impasse: if one denies that a statement that adds things together is a definition, will there be a definition of anything that is not simple but consists of things linked together? For it is clear that it would have to be defined by way of addition. I mean, for example, that there is a nose and there is being-squashed-in, and there is also snubness that means something made out of these two, this one in that one, and it is not incidental that the being-squashed-in or the snubness is an attribute of the nose, but in its own right, nor is it in the way that paleness is in Callias, or in a human being, because Callias, who is incidentally a human being, is pale, but in the way that maleness

1030b 20

---

[7] That is, since a quantity (say) is a being only in a qualified sense (with an addition), it is a being in less than the full sense (with a subtraction).

[8] See the first paragraph of Bk. IV, Ch. 2.

is in an animal or equality in an amount, and all those things that are said to belong to something in virtue of themselves. And these are those things in which there is present either the meaning or the name of that of which each is an attribute, and which cannot be displayed separately, as paleness can be displayed without a human being, but femaleness cannot be displayed without an animal. Therefore, there is either no what-it-is-for-them-to-be and definition of these things, or, if there is, it is in a different way, just as we have said.

But there is also a different impasse about them. For if a snub nose and a squashed-in nose are the same thing, then snub and squashed-in will be the same thing; and if they are not, on account of the impossibility of saying snubness without the thing of which it is an attribute in its own right (since snubness is squashed-in-ness in a nose), then it is either not possible to say snub nose, or else the same thing will have been said twice, squashed-in nose nose (since the nose that is snub would be a nose that is a squashed-in nose), for which reason it is absurd that a what-it-is-for-it-to-be should belong to such things, and if it did, it would be infinite, since in a snub nose there will always be another nose present.[9]

Accordingly, it is clear that definition belongs to thinghood alone. For if it belonged also to the other ways of attributing being, it would have to be by way of adding things together, for instance if there were a definition of the odd, since it is not without a number, nor is there a female without an animal (and by "by way of addition" I mean those statements in which one turns out to have said the same thing twice, as in these instances). But if this is true, there will be no definitions of linked things[10] either, such as of odd number, though this goes unnoticed because our statements are not articulated in a precise way.

1030b 30

1031a

---

[9] It takes some contortion to see the meaning of snub nose as infinite, but it may be worth the effort. If one simply substitutes "squashed-in nose" for the word snub, then the phrase contains "nose" twice, but only twice; but if one focuses only on the fact that snub contains an implicit reference to a nose, then snub nose becomes snub (nose) nose, which in turn becomes snub [nose (nose)] nose, and so on. From this perspective the phrase snub nose is infinite and therefore unbounded, and so cannot have a definition, which is a boundary.

[10] Chapters 4 and 5 may be said to deal with two kinds of "linked" things, as seen in the examples of a pale human being and a female human being. The *ti ēn einai* is what something keeps on being, in order to be at all, and so cannot include paleness, which a human being might happen to lose, or femaleness, which is not necessary in order for someone to be a human being. A human being must have some sort of

But if there are definitions of these, it is either in another way or else, just as was said, one must say that the definition and what it is for something to be are present in more than one way. Therefore in one sense there will not be a definition of anything, nor a what-it-is-for-something-to-be present in anything, except of and in independent things, but in another sense there will be. That, then, the definition is a statement of what it is for something to be, and that the what-it-is-for-something-to-be belongs to independent things alone, or else to them most of all, primarily, and simply, is clear.

1031a 10

Chapter 6   But one must investigate whether each thing is the same as, or different from, what it keeps on being in order for it to be. For this will contribute in a certain way toward the investigation about thinghood, since it is the case both that each thing seems to be nothing other than its own thinghood, and that what it is for it to be is said to be the thinghood of each thing. Now in the case of things meant as incidental, they would seem to be different, as a pale person is different from being-a-pale-person. (For if they were the same, then also being-a-human-being and being-a-pale-human-being would be the same; for a pale human being is just the same thing as a human being, as people say, so that being-a-pale-human-being and being-a-human-being would also be the same. Or does it not necessarily follow that they are the same in those instances that involve incidental attributes, since the extreme terms are not equated with the middle term in the same way?[11] But then perhaps this might seem to follow, that a pair of incidental extreme terms would become the same, such as being-pale and being-cultivated, but this doesn't seem to be so.[12])
But in the case of things attributed in virtue of themselves—such as if, for instance, there are some independent things than which no

1031a 20

---

generative capacity, but either sort, and even a defective instance of it, would still be sufficient.

[11] Being-a-pale-human is taken as identical with a pale human by assumption, while a pale human is taken as just the same as a human by loose ordinary speech, which doesn't discriminate between incidental and necessary attributes.

[12] If there was no syllogism in the previous case because of a lack of parallelism of the premises, would it help to make them both predications of incidental attributes? Obviously not, so one can now conclude than an incidental compound such as pale-human-being is not part of what it is to be anything.

1031a 30   other independent things or natures are more primary, of the sort that
some people say the forms are—is it necessary that *they* be the same
as what it is for them to be? For if the good itself and being-good
were different, or animal-itself and being-an-animal, or being itself

1031b   and being-being, then there would be other independent things and
natures and forms besides the ones that are spoken of, and those other
kinds of thinghood would be more primary, if what it is for something
to be is its thinghood. And if they are detached from one another,
there would be no knowledge of the forms, and what it is for things
to be would not have being. (By "being detached," I mean if being-
good does not belong to the good itself, and the good that *is* does not
belong to being-good.) For there is knowledge of anything only when
we recognize what it is for it to be, and the same thing that holds true
with the good holds also with the other things, so that if being-good
is not good, then being-being does not have being and being-one is
not a unity; and likewise, either for all things or for none, the what-
it-is-to-be has being, so that if being-being does not have being, then

1031b 10   neither do any of the others. What's more, that to which being-good
does not belong is not good. Therefore the good and being-good must
be one thing, and so too the beautiful and being-beautiful, and all
those things that are not attributed in virtue of anything else but in
virtue of themselves and primarily. For this would be sufficient if it
were granted, even if there were no forms, and perhaps even more so
if there are forms. (And at the same time it is also clear that, if there
are forms of the sort that some people say there are, thinghood will
not be an underlying thing; for these must be the thinghood of things,
but not as underlying things, since the other things will be by means
of participating in them.¹³)

So by these arguments, each thing itself and what it is for it to be are

1031b 20   one and the same, in a way that is not incidental, and this follows also
because knowing each of them is just this: to know what it is for it to be.

---

¹³ It was concluded at 1029a 26–30 that thinghood must refer to some sort of
underlying thing, but also to something that is separate and a *this,* and so must be the
form (or the composite) rather than the material. But if the form is only participated
in, or shared in, it cannot properly be said to underlie the particulars. This is not a
rejection of the idea of forms, but a fine tuning of one. Aristotle will argue that the
form must be at work upon the material in a way that makes the particular thing what
it is (Bk. VII, Ch. 17, and all of Bks. VIII and IX); the form therefore underlies the thing
in an active, causal way, and is not passively participated in.

Therefore, directly from the setting out of the question, it is necessary that both of them be some one thing. (But of what is attributed as incidental, such as cultivated or pale, it is not true to say that it and what it is for it to be are the same thing, because of its double meaning, since both that to which it is attributed, and the attribute, are pale, so that in one sense it and what it is for it to be are the same thing, but in another sense they are not, for being pale is not the same as being human, or even as being a pale human, but it is the same as being a certain attribute.) The absurdity [of distinguishing the form from what it is for something to be] would be evident if one were to put a name on each kind of thing there is for something to be, for then there would be another one besides that one; for example, what it is to be a horse would be different from what it is to be what it is to be a horse. So even in the first place, what prevents some things from immediately being what it is for them to be, especially since thinghood is what it is for something to be? But in fact, not only are they one thing, but even the very statement of them is the same, as is clear from what has been said; for it is in no incidental way that one and being-one are one. What's more, if they were different, they would be part of an infinite succession; for what it is for one to be would be one thing, and the one would be another, so that there would be the same argument in the case of each of these.

1031b 30

1032a

So it is clear that in the case of primary things, attributed in virtue of themselves, the being of each of them and each of them itself are one thing; as for the sophistical refutations directed at this thesis, it is obvious that they are resolved by the same resolution that applies to the question whether Socrates and being-Socrates are the same, since there is no difference either in the grounds on which one might raise the question, or in those on which one would succeed in resolving it.[14] In what way, then, what it is for something to be is the same as each thing, and in what way it is not, have been said.

1032a 10

---

[14] Some commentators take this to mean that, just as Socrates and his soul are identical, so too are the form and what it is for it to be; others take it to mean that, since what it is to be Socrates is identical with humanness in general, and not with him in particular, what it is to be Socrates is obviously different from Socrates, but on grounds that lead to an opposite conclusion in the case of the forms. It is perhaps best to understand Aristotle as saying that the application of the question to composite particulars needs more thorough examination, which Ch. 7–9 provide.

Chapter 7 Of the things that come into being, some come about by nature, some by art, and some as a result of chance, but everything that comes into being becomes something, from something, and by the action of something; by what it becomes, I mean something in accordance with any of the ways of attributing being, for it comes to be either a this, or how-much, or of-what-sort, or where it is. And natural comings-into-being are those of which the origin is from nature, and that out of which they come to be is what we call material, that by the action of which is any of the natural beings, and what they become is either a human being or a plant or anything else of that sort, which

1032a 20   in fact we most of all say are independent things—and all things that come into being by either nature or art have material, for each of them is capable of being and of not being, and this potentiality is the material in each—and in general, that out of which they come is a nature and that toward which they come to be is a nature (for the thing that comes into being, such as a plant or an animal, has a nature), and that by the action of which they come to be is the nature that is meant in the sense of the form and is the same in form as what comes into being (though it is in another, since a human being begets a human being).

It is in this way, then, that things come into being that come to be by nature, but the other things that come into being are called products. And all products result from art, or from an aptitude, or from thinking. But some of these things come about also just on their

1032a 30   own and by chance, in much the same way as happens among things that come into being as a result of nature; for there too, some of the same things that come into being from seeds are also produced without

1032b   seeds.[15] These must be looked into later, but it is from art that all those things come into being whose forms are in the soul (and by form I mean what it is for them to be, and their primary thinghood). For in a certain way, the same form belongs even to contrary things, since the thinghood of something lacking is the thinghood opposite to it,

---

[15] To all naked-eye appearances, a number of species produce offspring "spontaneously," contrary to the usual course of nature. These include mistletoe, among plants, and a variety of shellfish, among plantlike animals, as well as worms that arise in mud or decaying flesh. Something in the environment appears to play the role of the female generative fluid, and some part of the living thing substitutes for the semen. See *Generation of Animals* 762a 37–763b 16. The discovery, by means of the microscope, of the egg and sperm cells, brings these anomalies into Aristotle's account of the normal course of nature.

as health is of disease, for it is by the absence of health that there is disease, while health is a pattern and a knowledge in the soul. And something healthy comes into being when someone thinks in this way: since health is such-and-such, it is necessary, if something is to be healthy, that such-and-such be present, for instance uniformity, and if this is to be present there must be warmth,[16] and one goes on thinking continually in this way until one traces the series back to that which, at last, one is oneself capable of making. From that point on, the motion is then called production, namely the motion toward being-healthy. So it turns out that in a certain way health comes into being from health, and a house from a house, the one that has material coming from the one that is without material. For the medical art *is* the form of health, and the house-building art is the form of a house,[17] and by thinghood without material, I mean what it is for something to be.

1032b 10

So of the process of coming-into-being and the motion involved in it, one part should be called thinking and the other producing, the thinking starting from the source and from the form, and the producing starting from the completion of the thinking. And it is in a similar way too that each of the other steps in-between comes about. I mean, for example, that if someone is going to be healthy, he needs to be brought into a uniform condition. What, then, is it to be made uniform? Such-and-such, and this will be the case if he is made warm. And what is that? Such-and-such, which is present potentially, and this is already in one's power. So that which produces, and from which the motion of healing takes its origin, if it comes about by art, is the form in the soul, but if it comes about by chance, it is from whatever step it was that was the start of the producing for the one who produced it by art, as, in the doctoring in particular, the start was perhaps from the warming (and one produces this by rubbing). Accordingly, the warmth in the body is either part of health, or else something of a similar kind follows it, which is a part of health, or it follows by a number of steps; and the

1032b 20

---

[16] Aristotle compares the internal temperature of an animal to the sun's heat that ripens fruit and the cooking that makes food nourishing. See *Meteorology*, Bk. IV, Ch. 2–3, and *Generation of Animals* 743a 18–743b 18.

[17] We speak of the artist or artisan as a creator, but Aristotle understands the form to be at work upon the soul of the artist, who can then assist its transmission into material. In the famous example of the four causes of a statue, the sculptor is not even its "efficient" cause, except incidentally; the moving cause of a statue is the sculptor's art, a being-at-work of forms. (See above, 1013b 4–9.)

1032b 30

last of these is what produces the part, and is in this way itself a part, of health, or in the case of a house it is, say, stones, and so too in other things.[18] Therefore, just as is always said, coming into being would be impossible if there were nothing present beforehand.[19]

1033a

So it is clear that some part of the product will necessarily be present from the start, since the material is part of it (for this is a constituent in it and comes to be something). But this is not one of the things included in the articulation of a thing, is it? But we do state in both ways what bronze circles are, stating both that the material is bronze and that the form is a certain sort of shape, and this is the class into which it is primarily placed. So the bronze circle has its material in its articulation. But the material out of which some things come when they are generated is attributed to them not by that name but by one derived from it, as if a statue were not called stone but stony; but the human being grown healthy is not called by the name of that out of which he becomes healthy, and the reason is that he becomes healthy out of a deprivation and something underlying it, which is what we

1033a 10

call material (for instance, it's both a human being and a sick person that becomes healthy), though he is more so said to become healthy out of its lack, namely from being sick, than out of being human, and hence the healthy person is not called sick but human, and the human being is called healthy. But those things of which the lack is not apparent and has no name, such as the lack of any particular shape in bronze, or the lack of a house in bricks and lumber, seem to have things come into being out of them in the same way that the person in the earlier example became something out of being sick. And for this reason, just as there that out of which he comes to be is not attributed to him, here too the statue is not called wood but by the derivative wooden instead of wood, or brazen instead of brass, or stony instead of stone, and a

1033a 20

house is not said to be bricks but of bricks, since if one were to look very carefully, one would not say simply, either that the statue comes

---

[18] No chance heap of stones spontaneously turns into a shelter, but if it did, the process would start just where the housebuilder stops planning and starts producing. On the other hand, any friction that warms the body might happen to produce health; the rubbing is already part of health since it starts a sequence of events in which health is brought about. There is no implication that the two cases are alike.

[19] "Nothing comes from nothing" is true, but can easily lead to false conclusions. See *Physics*, Bk. I, Ch. 8.

into being out of wood, or the house out of bricks, since they have to
change and not remain what they are, in order for something to come
into being. This is why we speak in this way.[20]

Chapter 8   Now since what comes into being comes about by the
action of something (and by this I mean that from which the source of
its generation comes), and out of something (and let this be not the lack
but the material, since we have already distinguished the way in which
we mean this), and becomes something (and this is either a sphere or a
circle or whatever it might be in other cases), just as one does not make
the underlying thing, the bronze, so too one does not make the sphere,
except in the incidental sense that the bronze sphere is a sphere, and   1033a 30
one makes that. For to make a *this* is to make a *this* out of the whole of
what underlies it. (I mean that making the bronze round is not making
the "round" or the sphere but something different, such that this form
is in something else; for if one made the form, one would make it out
of some other thing, since that was assumed, in some such way as one   1033b
makes a bronze sphere, in the sense that out of this, which is bronze,
one makes this, which is a sphere.) So if one were also to make the
underlying thing itself, it is clear that one would make it in the same
way, and the comings-into-being would march off to infinity.

Therefore it is clear that the form, or whatever one ought to call
the shapeliness that is worked into the perceptible thing, does not
come into being, and that coming-into-being does not even pertain to
it, or to what it is for something to be (for this is what comes to be
in something else, by art or by nature, or by some capacity). But one
makes the bronze be a sphere, for one makes it out of bronze *and* out
of the sphere, since one brings the form into this material, and it is   1033b 10
this that is a bronze sphere. But of being-a-sphere, if there were to be
a coming-into-being of it at all, it would be something made out of
something. For the thing that comes into being will always have to be
divisible, and be not only this but also that—I mean, not only form
but also material. So if a sphere is a figure that is equidistant from the

---

[20] Aristotle's point is not to explain these verbal distinctions, but to explain them
away. The material of anything that comes into being is part of its very articulation, of
what might seem to be its formal aspect, even when we fail to notice it (by focusing
on the lack to the exclusion of the underlying thing that had the lack), have no name
for it, or alter its name to reflect its own altered condition.

center, one part of this would be that in which one puts something, and the other that which one puts in it, and the whole would be a product that has come into being, analogous to the bronze sphere. So it is clear from what has been said that what is spoken of as form or thinghood does not come into being, but the composite whole that is named in consequence of this does come into being; and it is clear that there is material present in everything that comes into being, so that it is not only this but also that.

1033b 20   But is there, then, some sphere apart from the ones around us, or a house apart from bricks? Or would there not even be any coming-into-being if the form were a *this* in that way, but does it instead indicate a certain kind, without being *this* and determinate? So it is rather the case that one makes or begets a certain kind of thing out of some *this,* and when it has been generated it is this-thing-of-this-kind. And the whole *this,* Callias or Socrates, is just like the sphere that is this bronze one here, while human being or animal is just like bronze sphere in general. Therefore it is clear that the causal responsibility attributed to the forms, in the sense that some people are in the habit of speaking of the forms, as if they are certain things apart from the particulars, is of no use, at least in relation to coming into being and independent things; nor would they be, for the sake of these things at least, independent

1033b 30   things in their own right.[21] With some things, in fact, it is even obvious that what does the generating is something of the same sort as the thing generated, though it is not the same one nor are they one in number, but they are one in kind, namely among natural things— for it is a human being that begets a human being—unless something is generated contrary to nature, such as a horse that's half donkey. (And even these cases are of a similar sort, for the narrowest class that

1034a   would be common to a horse and a donkey does not have a name, but it would presumably be of both kinds, as the half-donkey is.[22]) So it is clear that there is no need to go to the trouble of providing a form as a pattern (since they would have looked for it most of all among things

---

[21] Note that this heavily qualified claim rejects only a certain way of arguing for separate forms. Aristotle has already said that the form is a *this* and separate (1029a 27–30 and 1017b 25–28), and will make clear the sense in which *he* means this as the inquiry unfolds.

[22] Greek had no separate word for the hybrid, which we call a mule. It was well known that all mules are sterile. (See *Generation of Animals,* Bk. II, Ch. 8.)

generated by nature, for these most of all are independent things), but the begetter is sufficient to produce the things that come into being, and is responsible for the form's being in the material. But the whole, this particular form in these particular bones and flesh, is already Callias or Socrates; and they are different on account of their material (since it is different), but they are the same in form (for the form is indivisible).[23]

Chapter 9   One might be at a loss to explain why some things come about both by art and just on their own, such as health, while others, for instance a house, do not. But the reason is that in some things, the material that starts off the coming-into-being in the producing or becoming-something of what results from art, in which some part of the thing produced is present, is of such a sort as to be set in motion either by itself or not, and some of this is able to be moved by itself in a particular way, but some is not capable of it; for many things are capable of being moved by themselves, but not in some particular way, say dancing. So those things whose material is of this sort, such as rocks, are not capable of being moved in a particular way except by something else, but in another particular way, yes they are. This is why some things would not *be* without someone who has an art, while others would, since the latter will be set in motion by those things that do not have the art, since they can be moved by other things that do not have art or by a part of themselves.

And it is clear also from what has been said that in a certain way everything comes into being from something that shares its name, just as the things do that are by nature (for instance a house comes from a house, insofar as it comes about by the action of an intelligence, since its form *is* the art by which it is built), or from a part of such a thing, or from something having some such part, unless it comes into being incidentally; for the primary thing responsible for making something is, in virtue of itself, a part of what is made. For it is the heat in a motion that produces the heat in the body, and this is either health, or part of it, or some part of health follows along with it, or health itself does; hence it is also said to produce health, because it produces that which

1034a 10

1034a 20

1034a 30

---

23 To the question at the end of Ch. 6, this result would seem to imply the answer: no, Socrates is not the same as what it is for him to be, since then he and Callias would be the same. But perhaps, primarily and in virtue of themselves, they are the same. The topic comes up again in Ch. 11.

health follows along with and is an attribute of.

Therefore, just as in demonstrative reasoning, thinghood is the source of everything; for syllogisms come from what something is,[24] while here generations do. And the things composed by nature are in a condition similar to those produced by art. For the seed produces just as works of art are produced, since it contains the form in potency, and that from which the seed comes is, in some respect, a thing of the same name (unless there is a defect, which is why a half-donkey does not come from a half-donkey), for one needn't look for all things to be just the same as a human being from a human being, since it is also true that a woman comes from a man. And those things that come about on their own in nature come into being just as in the case of art, being those of which the material is capable of being moved by itself in the same motion which the seed sets moving; all those whose material is not of that sort are incapable of coming into being in any other way than from things of the same kind.

And it is not only about thinghood that the argument shows that the form does not come into being, but in the same way, the argument concerns in common all the primary things, such as how much something is, and of what sort, and the other ways of attributing being. For just as the bronze sphere comes into being, but not sphere or bronze, and also in the case of the bronze if it does come into being (for always it is necessary that material and form be present beforehand), so too in the case of what something is, and of what sort it is, and how much, and similarly with the rest of the ways of attributing being, for it is not the this-sort that comes into being, but a wooden thing of this sort, and not the so-much but so much wood or an animal that is so big. But what is to be understood from these considerations as peculiar to an independent thing is that a different independent thing that is fully at work, and that makes it, must be present beforehand, such as an animal if it is an animal that comes into being; but with what is of this sort or so much, this is not necessary, but only something that is potentially each.

1034b

1034b 10

1034b 20   **Chapter 10** Now since the definition is a statement and every

---

[24] This is the ultimate presupposition of any reasoning. See *Posterior Analytics* 90b 31–33.

statement has parts, and as the statement is related to the thing defined, so too is the part of the statement related to the part of the thing, the difficulty immediately arises, whether it is or is not necessary for an articulation of the parts to be present in that of the whole. For with some things they obviously are included, but with others they obviously are not. For the articulation of the circle does not contain that of its segments, but the articulation of a syllable does contain a statement of its letters, even though the circle is divided into its segments just as the syllable is divided into its letters. And further, if the parts precede the whole, while the acute angle is part of the right angle and a finger is part of an animal, the acute angle would take precedence over the right angle, and the finger over the human being. But it is those others that seem to be more primary; for in an account of them, the parts are explained by means of the wholes, and in regard to which of them can have being without the others, the wholes are more primary.

1034b 30

Or else a part is meant in more than one way, one of which is in the sense of what measures how much something is—but let this sense be left aside; but those things of which thinghood consists in the sense of parts are what one needs to examine. So if material is one thing, and form another, and another is what is made of these, and an independent thing is both the material and the form as well as what is made of them, there is both a sense in which the material is said to be part of something, and a sense in which it is not, but the parts are those things of which the articulation of its form consists. For instance, flesh is not part of being-squashed-in (since this is the material upon which it comes to be), but it is a part of snubness; and of the composite whole of a statue, bronze is a part, but not of the statue described as a form (for what one must state is the form, and each thing must be described insofar as it has form, while what is simply material as such should never be stated[25]). This is why the articulation of the circle does not contain that of its segments, while that of a syllable does contain a statement of its letters; for the letters are not material but are parts of the form, while the segments are parts in the sense of material upon which the circle is begotten, yet even so they are nearer to the form than is the bronze, in those cases in which roundness becomes present

1035a

1035a 10

[25] The material is always to some extent incidental to what a thing is. Aristotle says in *Physics* Bk. II, Ch. 9, that a saw has to be made of iron, but at 1044a 27–29 below he says merely that some materials, such as wood or wool, would not make it a saw.

in bronze. And there is a sense in which not even all the letters will be included in the articulation of a syllable, such as the particular ones in a wax tablet, or the ones uttered in the air, since these also already count as part of the syllable in the sense of perceptible material. For even if a line when it is divided passes away into its halves, or a human being into bones, connective tissue, and flesh, it does not follow that

1035a 20   for this reason they are made out of those things in such a way that they are parts of the thinghood of them, but only that they are made out of them as material, and as parts of the composite whole; but this does not go so far as to make them parts of the form or of that which the articulation is about, and for just that reason they are not in the articulations.

So for some things, the articulation of such parts will be included, but for others it ought not to be included, when the articulation is not of the thing as taking in its material with it; for that is why some things are made out of these parts as sources into which they pass away, while some are not. So as many things as take in the form and material together, such as snubness or a bronze circle, do pass away into them and the material is part of them, but as many as do not take in the material with them, but are without material, the articulations

1035a 30   of which refer to the form alone, do not pass away, either not at all or at least not in that way; and so of the former things, these materials are sources and parts, but are neither parts nor sources of the form. And it is for this reason that a clay statue passes away into clay, a sphere into bronze, and Callias into flesh and bones, and even a circle into its segments, since there is a kind of circle which takes in its material

1035b   along with it; for circle is meant ambiguously, both simply and as a particular one, since there is no special word for the particular ones.

So even now, the truth of the matter has been stated, but still let us speak still more clearly, taking it up again. For those parts that belong to the articulation of a thing, and into which that articulation is divided, are more primary than it is, either all of them or some of them; and the articulation of the right angle does not have the articulation of the acute angle as a division of it, but that of the acute does have that of the right as a division, for someone defining the acute angle makes use of the right angle, since the acute is that which is less than a right

1035b 10   angle. And the circle and semicircle are also similarly related, since the semicircle is defined by means of the circle, and also the finger by means of the whole, since a finger is a certain sort of part of a human being. And so all those things that are parts in the sense of material, and into which something divides up as into material, are derivative from the whole; but either all or some of those that are parts in the sense of belonging to the articulation and to the thinghood that is disclosed in the articulation, are more primary than it.

And since the soul of an animal (for this is the thinghood of an ensouled thing) is its thinghood as disclosed in speech, and its form, and what it is for a certain sort of body to be (at any rate, each part of it, if it is defined well, will not be defined without its activity, which will not belong to it without perception[26]), either all or some of the parts of the soul are more primary than the whole animal as a composite, and similarly with each particular kind, but the body and its parts are derivative from the thinghood in this sense, and it is not the thinghood but the composite whole that divides up into these as into material. Now in a sense these are more primary than the composite, but in a sense they are not (for the parts of the body are not capable even of being when they are separated, for something is not the finger of an animal when it is in any condition at all, but that of a corpse is a finger in only an ambiguous sense), and some of them are of equal primacy with the whole, those that are governing and in which the articulation and thinghood primarily are, whether this is (say) the heart or the brain, for it makes no difference which part is of that sort. But a human being or horse in general, and the things that are in this way after the manner of particulars, but universally, are not thinghood but a certain kind of composite of such-and-such an articulation with such-and-such material, understood universally, while the particular, composed of ultimate material, is already Socrates, and similarly in other cases.

So a part belongs either to the form (and by form I mean what it is for something to be) or to the composite of material and form, or to the material itself. But the parts of a thing's articulation belong only to the form, and the articulation is of the universal; for being a circle and a circle, or being a soul and the soul, are the same thing. But of the composite there is already no definition, for instance of this circle here,

1035b 20

1035b 30

1036a

---

26 The capacity that distinguishes animals from plants is perception, which in some way permeates everything an animal does. An animal is not a plant plus sense organs, but a genuine whole, and perception is not one function among many but a governing one.

or of any of the particular ones, either perceptible or intelligible—by intelligible ones I mean such ones as belong to mathematics, and by perceptible ones, such ones as are bronze or wooden—but we know them directly by the contemplative intellect or by sense perception, and once these fall away from an active exercise, it is not clear whether they have being or not, but they are always described and known by means of a universal articulation.[27] But the material is not known in its own

1036a 10   right. And one sort of material is perceptible, the other intelligible, the perceptible, for example, bronze or wood, or any movable material, while the intelligible is that which is present in perceptible things, taken not as perceptible, as for example mathematical things are.

So how it is with whole and part, and with preceding and following, has been said; and it is necessary to approach the question, whenever someone asks whether a right angle or a circle or an animal is more primary than the parts into which it is divided and out of which it is made, by saying that it is not simply one or the other. For if the soul is the animal or ensouled thing, or each of them is its own soul, and a circle is being-a-circle, and a right angle is being-a-right-angle or the thinghood of the right angle, then any of them must be said to be

1036a 20   derivative from something, such as from the things in its articulation, and it is also derivative from its parts if the question is asked about a particular right angle (for this is true both of the one with material, the bronze right angle, and of the one that is in particular lines), but the one without material is derivative from the things in its articulation, but more primary than the parts in each particular, but it is not something that one can state simply. And if the soul is a different thing and is not the animal, in that sense too one must say that some things are more primary than it and that some are not, as has been said.

Chapter 11   One might reasonably be confused about what sort of things are parts of the form, and what sort are parts not of that but of the all-inclusive composite. And yet so long as this is not clear, it is not possible to define any particular thing, since the definition is of the universal and the form; so if it is not clear what sort of parts are

1036a 30   present in the manner of material and what sort not, neither will the

---

[27] This is a very important passage. It says that the universal is a reconstruction in speech of the form, while the form itself is present directly to the perceiving or contemplating soul.

articulation of the thing be evident. Now it seems clear for all those things that are obviously brought into being in materials different in form,[28] such as a circle in bronze or stone or wood, that the bronze or the wood does not in any way belong to the thinghood of the circle, because of its being separated from them; but nothing prevents those things that are not seen to be separated from being similar to the others, just as if all the circles one had seen were bronze, since nonetheless, the bronze would in no way belong to the form, though it would be difficult to subtract it in one's thinking. For example, the form of a human being always appears in flesh and bones and parts of that sort: are they then also parts of the form and of its articulation? Or are they not, but just material, though because humans are not brought into being in other materials we are unable to separate them?

1036b

But since this seems to be possible, while it is not clear when it is the case, some people are already at an impasse even about the circle and the triangle on the grounds that it doesn't seem right to define them by means of lines and continuous magnitude, but rather to speak of all these too as if they were similar to the flesh and bones of a human being and the bronze and wood of a statue; and these people trace all things back to numbers, and say that the articulation of a line is that of the number two. And among those who speak about forms, some say the number two is the line itself, others that it is the form *of* the line, for with some things they say that the form and that of which it is the form are the same (such as the number two and the form of twoness), but that this does not extend as far as to the line.[29] So it turns out that there is one form of many things of which the form seems different (the very thing that also turns out to be so for the Pythagoreans), and also that it is possible to make out of everything one thing that is the form itself, and make the others not be forms; and yet in this way all things would be one.

1036b 10

1036b 20

---

[28] One might translate *eidos* here as "kind" or "species," but Aristotle is being very precise. As he says above at 1036a 8–9, material as such is unknowable; we know it only by *its* form.

[29] The original identification of the line with the number two is Pythagorean; two points determine a line, which stretches in two directions, etc. The distinction about the forms relates to an understanding that even among the forms there must be something that plays the role of material, to make the line different from two itself.

That, then, the things pertaining to definitions hold a certain impasse, and why they do, have been said; and that is why tracing everything back in this way, and taking away the material, is overly fastidious, for presumably some things are such-and-such *in* such-and-such, or such-and-such in such-and-such a condition. And the analogy that Socrates the younger used to make in regard to an animal is not a good one, for it leads away from the truth and makes one assume that it is possible for there to be a human being without parts, just as there can be a circle without bronze. But the animal is not like the circle, since it is something perceptible, and it cannot be defined leaving out

1036b 30   motion, nor for that reason without parts that are disposed in a certain way. For it is not a hand of any sort that is part of a human being, but one capable of accomplishing its work, and therefore being ensouled; and what is not ensouled is no part of it.

But then in the case of mathematical things, why aren't the articulations of the parts included in those of the wholes, such as that of the semicircles in that of the circle? For these are not perceptible things. Or does it make no difference? For there will be material even of some

1037a   things that are not perceptible, since in everything that is not what it is for something to be, nor a form itself just as itself, but a *this*, there is a certain material. So these parts will not belong to the circle understood universally, but will belong to the particular circles, just as was said before; for there is not only perceptible but also intelligible material.

And it is clear too that the soul is the primary independent thing, while the body is material, and the human being or animal in general is what is made of both, understood universally; and if it is also true that the soul of Socrates *is* Socrates, then names such as Socrates or Coriscus have two meanings (for some people mean by them a soul, but others the composite), but if Socrates is simply this

1037a 10   soul plus this body, then the particular is just like the universal. But whether there is some other material besides that which belongs to independent things of this kind, and whether one needs to look for some other kind of independent thing such as numbers or anything of that sort, one must consider later. For it is for the sake of this that we are also trying to mark out the boundaries of those independent things that are perceptible, since in a certain way the study of perceptible beings is the work of the study of nature, that is, of second philosophy; for the one who studies nature must know not only about the material but also about what is disclosed in speech, and even more so. And in the case of definitions, one must consider later in what way the things in an articulation are parts, and what makes an articulation that is a definition be one. (For it is clear that the thing defined is one, but by means of what is it one, since it has parts?)[30]

1037a 20

What, then, the what-it-keeps-on-being-in-order-to-be-at-all of something is, and in what way it is what a thing itself is in virtue of itself, have been stated in a general way that applies to everything, as has the reason why the articulation of what it is for some things to be has in it the parts of the thing defined while that of other things does not, and that the parts in the sense of material are not included in the articulation of the thinghood of a thing—for they are not parts of that kind of thinghood, but only of the composite whole, and of this in a certain way there is and there isn't an articulation; for there is none with the material (since this is indeterminate), but there is one in virtue of the primary thinghood of the composite, as for instance the articulation of the soul of a human being. For the thinghood of a composite is the form that is in it, and the whole that is made out of that and the material is called an independent thing (such an indwelling form is squashed-in-ness, for a snub nose, or snubness, is made out of that and a nose)—though in the composite independent thing, such as a snub nose or Callias, the material is also included. And it has been said that each thing and what it is for it to be are in some cases the same, as with the primary independent things (and by primary I mean what is not attributed to anything else so as to be in another thing that underlies it as material), but as many things as are material or as take in material along with them, are not the same as what it is for them to be, nor is anything that is one in an incidental way, such as Socrates and being cultivated,

1037a 30

1037b

---

[30] The question how the parts make one whole is resolved in Bk. VIII. That resolution also implies that Socrates is not a soul plus a body, but a unity of the two in a way that is attributable to the soul. So the best answer to the question whether Socrates is the same as what it is for Socrates to be, seems to be yes, though Aristotle leaves it with a yes-and-no at the end of this chapter. The question whether there are independent things besides the perceptible ones is resolved in Bk. XII. That is the culmination of first philosophy, to which the whole analysis of the being of perceptible things in Bks. VII–IX is subordinate. What is first for us is not first in itself.

since they are only incidentally the same as what it is for them to be.[31]

Chapter 12   Let us now speak first about definition, to the extent that it is not spoken about in the writings on the analytic art,[32] for the impasse that is stated in those discussions will contribute to the discussions about thinghood. The impasse I mean is this: why in the world oneness belongs to the articulation that we call a definition, such as the definition of a human being as a two-footed animal, for let that be its articulation.[33] Why is this one thing and not many, an animal and two footed? For in the case of human being and pale, they are many whenever the latter is not present in the former, but one when it is present and the underlying thing, the human being, is affected in a certain way (for then they become one thing and are a pale human being); here, though, one of them does not have any part in the other, since the general class does not seem to have a share in its specific differences (for it would have a share in contrary things at the same time, since those differences by which the genus is differentiated are contraries). And even if it does share in them, the same argument returns, so long as its specific differences are many, such as footed, two footed, featherless. For why are *these* one and not many? For it is not because they are all present in something, since in that way there will be a oneness of everything in it. But what is necessary is that there be a oneness of as many things as are in its definition, for the definition is a certain kind of articulation that is one and belongs to an independent thing, so that it has to be an articulation that belongs to something that is one; for an independent thing means something that is one and a *this,* as we assert.

1037b 10

1037b 20

---

[31] Two inapplicable passages have crept into the manuscript texts of this paragraph. One that gives curvature as an example of a primary being (presumably because it is more general than snubness and doesn't refer to material, though it is the human soul that is the primary being to which snub noses are incidental) is rejected by Jaeger in the most recent Oxford text. The other, that inserts an irrelevant reminder that snub nose has nose in it twice, is rejected by Ross in his earlier Oxford text. Neither passage belongs here.

[32] This is the collection of writings on logic sometimes called the *Organon.* The reference here is to *Posterior Analytics,* Bk. II, Ch. 3–10 and 13. But logic has no light to cast on the question about to be raised.

[33] On the possible non-silliness of this definition, see the footnote to 1006a 32.

And one ought first to examine the definitions that result from divisions.[34] For there is nothing in the definition other than the general class that is mentioned first and specific differences within it; each of the other general classes is the first one with the differences taken in along with it, such as, first, animal, next, two-footed animal, then, featherless two-footed animal, and it is said in the same way through even more steps. But it makes no difference at all whether it is stated by means of many steps or few, and therefore not even whether by means of a few or by means of two; and of the two, one is the specific difference and the other is the general class, as with two-footed animal, animal is the genus and the other part the specific difference. So if the genus taken simply does not have being apart from the species which belong to the genus, or if it does have being but as material (for the voice is a genus and material, and its differentiations make forms and letters out of that), it is clear that a definition is an articulation consisting of specific differences.[35]

1037b 30

1038a

But surely it is necessary also to divide the difference into its differences; for instance, provided-with-feet is a difference belonging to animal, and next one must recognize the difference within animal-provided-with-feet insofar as it is provided with feet, so that one ought not to say that of what is provided with feet, one sort is feathered and the other featherless, if one is to state things properly (for one would do this rather through ineptness), but instead that one sort is cloven footed and the other uncloven, since these are difference that belong to a foot, cloven-footedness being a certain kind of footedness. And one wants to go on continually in this way until one gets to things that have no differences; and then there will be just as many kinds of foot as there are specific differences, and the kinds of animal-provided-with-feet will be equal in number to the differences. So if that is the way these things are, it is clear that the difference that brings the statement to completion will be the thinghood of the thing and its definition, if indeed one ought

1038a 10

1038a 20

---

[34] This method is exhibited in Plato's *Sophist,* with a first example at 218B–221A. One starts with a general class wide enough to include the thing being defined, then keeps dividing it in two until a last class just fits. As Aristotle points out, one need not report anything but one last general class and the specific difference that restricts it to the nature defined.

[35] This first conclusion shows that the genus is a subordinate part in the definition, so that the question of its unity becomes, what makes the differences one?

not to repeat the same things many times in definitions, since that is overly fastidious. But this does in fact happen, for whenever one says animal, provided with feet and two footed, one has said nothing other than animal having feet, having two feet; and if one divides this by an appropriate division, one will say the same things even more times, times equal in number to the divisions.

So if a difference comes into being out of a difference, the one that brings this to completion will be the form and the thinghood of the thing; but if a difference is brought in incidentally, such as if one were to divide what is provided with feet into one sort that is white and another sort that is black, there would be as many differences as cuts. Therefore it is clear that a definition is an articulation consisting of

1038a 30   differences, and arising out of the last of these when it is right. And this would be evident if one were to rearrange definitions of this sort, such as that of a human being, stating "animal, two footed, provided with feet," since "provided with feet" is superfluous after one has said "two footed." But there is no ordering in the thinghood of the thing; for how is one supposed to think of one thing as following and another preceding?[36] So concerning definitions that result from divisions, let this much be said first about what sort of things they in particular are.

1038b   **Chapter 13**   But since the investigation concerns thinghood, let us go back to that again. And just as the underlying thing, and what it is for something to be, and what is made out of these are said to be thinghood, so too is the universal. Now what has been said concerns two of these (for it concerns what it is for something to be, and of the underlying thing it has been said that what underlies has two senses, being either a *this*, in the sense that an animal underlies its attributes, or material, in the sense that it underlies its complete being-at-work), but it seems to some people that the universal is responsible for a thing most of all, and that the universal is a governing source, and for that reason let us go over this. For it seems to be impossible for any of

1038b 10   the things meant universally to be thinghood. For in the first place, the thinghood of each thing is what each is on its own, which does not

---

[36] That is, all those intermediate differences that are obviously unnecessary when stated out of order, must be simply unnecessary in the thinghood itself, when not subjected to the temporal ordering of speech.

belong to it by virtue of anything else, while the universal is a common property, since what is meant universally is what is of such a nature as to belong to more than one thing. Of which of them, then, will it be the thinghood? For it is of either all or none, and of all it cannot be; but if it were of one, that one would also be the others, since those things of which the thinghood is one, and for which what it is for them to be is one thing, are also themselves one thing.

Again, thinghood is what is not attributed to any underlying thing, but the universal is always attributed to some underlying thing. But could it be like this: that the universal does not admit of being thinghood in the same way as what it is for something to be, yet is included in this, the way animal is in human being or horse? In that case, it is clear that there would be some articulation of it. And it makes no difference if it is not the articulation of everything in the thinghood   1038b 20
of a thing, for this will nonetheless be the thinghood of something, as human being is of the human being in whom it is present, so that the same thing will turn out again to be the case, since the thinghood would belong to that kind, such as animal, in which it would be present as peculiar to it. And what's more, it is impossible and absurd that what is a *this* and an independent thing, if it is composed of anything, should have as a component something that is not an independent thing or a *this* but an of-such-a-sort; for what is not an independent thing and is an of-this-sort would be more primary than an independent thing and a *this*. But that is exactly what cannot be, for neither in articulation, nor in time, nor in their coming-to-be could attributes be more primary than an independent thing, for they would also be separate. And on top of this, there would be another independent thing present in Socrates, so that an independent thing would belong to two things. And in general   1038b 30
it follows that if human being, and whatever things are meant in that way, are thinghood, none of the things present in their articulations are the thinghood of anything, nor are they present separately from them, nor in anything else; I mean, for example, that there is not any "animal" besides the particular kinds, nor do any of the other components in their articulations have being separately.

So for those who pay attention, it is clear from these things that nothing that belongs to anything universally is thinghood, and that   1039a
none of the things attributed as common properties signifies a this, but only an of-this-sort. And if this were not so, among other consequences

there would also be the third man.[37] And it is clear in this way too: for it is impossible for an independent thing to consist of independent things present in it in a fully active way, since what is in that way actively two could never be actively one, though if it were two potentially, it would be one. (For example, what is double is made of two halves, but only potentially, since being a half in a fully active way separates it.) Therefore, if an independent thing is one, it will not be made of independent things present in it in that way, which Democritus says

1039a 10   correctly; for he says it is impossible for one to come to be out of two or two out of one, since he make the independent things uncuttable magnitudes. Accordingly, it is likewise clear that this will hold also in the case of number, if a number is an assemblage of units as is said by some people; for either the number two is not one thing, or there is no unit actively in it.

But there is an impasse. For if no independent thing can be made out of universals, because the universal signifies an of-this-sort but not a this, and no independent thing admits of being composed of active independent things, every independent thing would be not composed of parts, so that there could not be an articulation in speech of any independent thing. But surely it seems to everyone and has been said

1039a 20   from earliest times that a definition belongs to an independent thing either solely or most of all; but now it seems not to belong to this either. Therefore there will be no definition of anything; or in a certain way there will be and in a certain way there will not. And what is said will be more clear from things said later.

Chapter 14   But it is also clear from these same things what follows for those who say that the forms are independent things and separate, and at the same time make the form be a compound of a general class and its specific differences. For if the forms have being, and animal is in human being and in horse, it is either the same thing and one in number, or it is different; for it is clear that it is one in articulation, since

1039a 30   the one stating it goes through the same articulation in both cases. If, then, there is a human being itself that in virtue of itself is a *this* and separate, it must also consist of things, such as animal and two footed, that signify a *this* and are separate and independent things, so that this

---

[37] See the footnote to 990b 17.

will be true in particular of animal. And then, if what is one and the same, just as you are with yourself, is in horse and in human being, how will something be one that is in things that have being separately, and for what reason will this animal not be separate even from itself? Next, if it is to partake in two-footedness as well as in many-footedness, something impossible will follow, since contrary things will belong at the same time to the very thing that is one and a *this;* while if it does not partake in both, what is the way of it when one says that animal is two footed or footed? Perhaps they are just sitting there together, either touching or mixed, but all that is absurd.

1039b

But if animal is different in different species, then those things of which animal is the thinghood would be, in a manner of speaking, unbounded,[38] since it is not in an incidental way that human being contains animal. What's more, animal itself will be many things, since the animal in each species is its thinghood (for it is not attributed in virtue of anything else, and if it were, then human being would consist of *that,* and that would be its genus), and further, all the things of which human being consists would be forms, and so none could be the form of one thing but the thinghood of another, for that would be impossible; therefore there will be one animal-itself each among the things in the various kinds of animal. Yet what does that come from, and how does it consist of animal itself? Or how could that be animal, for which this thing in question, aside from animal itself, is itself the thinghood? Furthermore, in the case of perceptible things, not only these results follow, but others more absurd than these. So if it is impossible for things to be this way, it is clear that there are not forms of perceptible things in the way that some people say there are.[39]

1039b 10

Chapter 15   Now since the composite whole and its articulation are different kinds of thinghood (and I mean that one kind of thinghood in this sense is the articulation with the material taken in along with it, while the other is entirely the articulation), there is destruction of all those things that are called independent things in the former sense (since there is also coming into being), but of the articulation

1039b 20

---

[38] The point is not that there might be infinitely many kinds of animal, but that the nature of each kind would have an indeterminacy at its core.

[39] Note once again that Aristotle does not dispute that there are separate forms which are themselves independent things, but only *how* there are.

there is no destruction in the sense that it is passing away (for neither is there coming-into-being of it, since it is not being-a-house
that comes about but being-this-house), but they are or are not without coming to be or perishing, since it has been shown that no one
either generates or makes them.[40] And this is why there is no definition of nor demonstration about particular perceptible independent

1039b 30   things: because they contain material that is of such a nature that
they are capable of being or not being, which is why all the particulars among them are destructible. So if demonstration is of what
is necessary, and definition is suited to exact knowledge, while perceptible particulars do not admit of these, then just as knowledge is
not sometimes knowledge but sometimes ignorance, but it is opinion
that is of that sort, and in the same way it is neither demonstra-

1040a   tion nor definition, but rather opinion, that concerns what admits
of being otherwise, it is clear that there could be no definition or
demonstration of these things. For destructible things are unclear to
those who have knowledge of them when they pass out of sense
perception, and even though articulations of them are preserved in
the soul, there will not be a definition any longer, nor a demonstration. For this reason it is necessary, when one is making distinctions
aiming at a definition of any of the particulars, not to be unaware
that it is always subject to be annulled, since the thing cannot be defined.

But then neither can any form be defined, since they say that

1040a 10   the form is a particular and is separate; but it is necessary that an
articulation be composed of words, and the definer will not make
up a word (since it would be unknown), but the words must be
names given in common to everything, so that they must also belong
to something else. For example, if someone were to define you, he
would say "a skinny, pale animal," or something else that would also
belong to some other thing. And if someone were to claim that nothing
prevents all the parts separately from belonging to many things, but
together they belong only to this one, one must say, first, that they
do belong to both of two things, for instance, in the case of two-
footed animal, both to the animal and to the two-footed. (And in
the case of everlasting things it is necessary even that the parts be

---

[40] This was shown in the first half of Ch. 8.

more primary than the compound of them, and rather, in fact, that they even be separate, if human being is separate. For either nothing is separate or both the parts are; so if nothing is, there would be no genus apart from its species, while if anything is to be separate, the specific difference is too.) And next, one must say that the parts are more primary in being, and so they are not wiped out when the wholes are.

1040a 20

And next after this, if the forms consist of forms (for the things of which they consist are more uncompounded than they are), those things of which the form consists, such as animal and two footed, will still have to apply to many things. If they did not, how would they be known? For there would be a form which was incapable of being attributed to more than one thing, but that doesn't seem possible, but rather that every form is capable of being shared in. So just as was said, it escapes notice that there is an impossibility of defining among everlasting things, but most of all that this is the case with those that are unique, such as the sun or moon. For people miss the mark not only by adding things of a sort such that, if they were taken away, the sun would still be the sun, such as "going around the earth" or "hidden at night" (for if it were to stand still or shine at night it would no longer be the sun, but it would be absurd if it were not, since "the sun" signifies a certain independent thing), but also by including things that admit of applying to something else, such that, if another thing of that kind came into being, it would clearly be a sun; therefore the articulation is common, but the sun was understood to be among the particular things, as is Cleon or Socrates. Otherwise, why does none of them come out with a definition of a form? For it would become clear when one tried it that what is now being said is true.

1040a 30

1040b

Chapter 16 And it is clear that most of what seem to be independent things are potencies, not only the parts of animals (since none of them is separate, and when they are separated, even then they all have being only as material), but also earth and fire and air, since none of them is one, but just like a heap, until some one thing is ripened or born out of them. Now one might assume especially that, in things with souls, the parts that are intimately related to the parts of the soul could come to be of both sorts, since they have being both in full activity and in potency, by having sources of motion stemming from something

1040b 10

in the joints,[41] on account of which some animals live when they are divided up. But still, whenever the parts are one and continuous by nature, and not by force or by growing parasitically, they will all have being as potencies; for the exceptions are defective instances.

And since one is meant in just the same way as being, and the thinghood that belongs to what is one is also one, and those things of which the thinghood is one in number are one in number, it is clear that neither oneness nor being admits of being the thinghood of things, just as being-an-element or being-a-source could not be thinghood; instead, we are seeking what it is that is the source, in order that we may trace things back to what is more knowable in itself. Now among these, being and oneness are thinghood more so than are sourcehood and elementality and causality, but it is not at all even these, if indeed nothing else that is a common property is thinghood either; for thinghood belongs to nothing other than itself and that which has it, of which it is the thinghood. What's more, what is one could not be in more than one place at the same time, but a common property is present in more than one place; therefore it is clear that none of the things that are present as universals is separate, apart from particulars.

1040b 20

But those who speak about the forms in one way speak rightly in separating them, if they are independent things, but in another way say wrongly that what is one-applied-to-many is a form.[42] The reason is that they cannot give a complete account of *which* things of that sort are independent and indestructible things, apart from the particular and perceptible things, so they make them the same in kind as the destructible things (since we know these), human-being-itself or horse-itself, adding to the perceptible things the word "itself." And yet, even if we had not seen the stars, nevertheless I suppose there would have

1040b 30

1041a

---

41 Any change of place by or in an animal must start from a joint, where one side stays fixed for the motion to push off against. (See *Motion of Animals,* Ch. 1.) So it might seem that a severed part could live in isolation, so long as it included a joint.

42 In Plato's dialogues, Socrates will often say something like, "aren't we accustomed to speak of the many beautiful things but also, distinct from these, one form that is the beautiful itself?" This is the beginning, not the end, of thinking about forms. If one hastily decides that wherever there is a many that can in any way be spoken of as one, that one thing is a form, then the forms become classes or universals or common properties (*genē, ta katholou, koina*), which cannot also be separate and have causal responsibility for things.

been everlasting independent things besides the ones we knew, so that now too, even if we cannot say what they are, it is still presumably necessary that there be some. That, then, none of the things attributed universally is an independent thing, and that no independent thing is composed of independent things, is clear.

Chapter 17   But what one ought to say thinghood is, and of what sort it is, let us speak about again, as though making another start; for perhaps from these discussions there will also be clarity about that kind of thinghood that is separate from perceptible independent things. Now since thinghood is a certain kind of source and cause, one must go after it from that starting point. And the why of things is always sought after in this way: why one thing belongs to something else.[43] For to look for the reason why a cultivated human being is a cultivated human being is to seek either what was just said, why the human being is cultivated, or something else. Now why something is itself is not a quest after anything (for the *that* or the being-so has to be present all along as something evident—I mean, say, that the moon is eclipsed—but "because a thing is itself" is one formulation and one cause that fits all, why a human being is a human being or cultivated is cultivated, unless someone were to say that each thing is indivisible from itself, and that is what it is to be one; but this is common to everything and a shortcutting of the question). But one could search for the reason why a human being is a certain sort of animal. And in that case this is clear, that one is not searching for the reason why that which is a human being is a human being; therefore, one is inquiring why something is present as belonging to something. (That it is present has to be evident, for if that is not so, one is inquiring after nothing.) For example, "why does it thunder?" is, "why does noise come about in the clouds?," for thus it is one thing's belonging to another that is inquired after. Or why are these things here, say bricks and stones, a house? It is clear, then, that one is looking for what is responsible, which in some cases, as presumably with a house or a bed, is that for the sake of which it is, but in some cases it is that which first set the

1041a 10

1041a 20

1041a 30

---

[43] In *Posterior Analytics* II, Ch. 1–2, Aristotle says that all questions fall into four kinds: what is the case, why, whether something exists, and what something is. He argues that they all go back to the why, as a search for a middle term through which one thing belongs to another.

thing in motion, since this too is responsible for it. But while the latter sort of cause is looked for in cases of coming into being and destruction, the former applies even to the being of something.

But the thing in question escapes notice most of all in those cases in which one thing is not said to belong to another, as when the thing one is seeking is what a human being is, because one states it simply and does not distinguish that these things are this thing. But it is necessary to inquire by dividing things at the joints; and if one does not do this, it becomes a cross between inquiring after nothing and inquiring after something. But since it is necessary that the being of something hold on to and be present to something, it is clear that one is asking why the material is something; so, "why are these things here a house?"— because what it is to be a house belongs to them. And this thing here, this body holding on in this condition, is a human being. Therefore what is being sought is the responsible thing by means of which the material is something, and this is the form. Accordingly, it is clear that in the case of simple things, there is no process of inquiry or teaching, but a different way of questing after such things.

But then there is what is composed of something in such a way that the whole is one, in the manner not of a heap but of a syllable—and the syllable is not the letters, nor are B plus A the same as the syllable BA, any more than flesh is fire plus earth (for when they are decomposed, the wholes, such as flesh or a syllable, no longer are, but the letters, or the fire and earth, are); therefore there is something that is the syllable, not only the letters, the vowel and the consonant, but also something else, and the flesh is not only fire and earth, or the hot and the cold, but also something else. Now if that something else must necessarily either be an element or be made of elements, then if it is an element there will be the same argument again (since flesh would be made of this plus fire plus earth, and something else again, so that it goes on to infinity), but if it consists of an element, obviously it would consist not of one but of more than one, or else it would itself be that one, so that again in this case we will state the same argument as in the case of the flesh or the syllable. But it would seem that this something else is something, and is not an element, and is in fact responsible for the flesh's being this and the syllable's being that, and similarly too in the other cases. But this is the thinghood of each thing (for that is what is primarily responsible for the being of it)—and since some things are not independent things, but those that are independent things are put

1041b

1041b 10

1041b 20

together by nature and in accordance with nature,[44] it would seem   1041b 30
that it is this nature that is thinghood, which is not an element but a
source—but an element is that which something is divided into, being
present in it as material, such as the A and the B of the syllable.

---

[44] Products of art or craft have thinghood in a derivative sense, since they borrow their
materials from natural things, and do not maintain themselves by activity. Random
heaps are scarcely things at all. Attributes and properties belong to wholes, and parts
are potencies that contribute to the maintenance of wholes. What else is there? At
this stage of the inquiry in quest of being itself, Aristotle has cleared away everything
but plants, animals, and the ordered cosmos.

Book VIII (Book H)
Form and Being-at-Work[1]

Chapter 1 Now one ought to reckon up the results of what has been said, and, putting them all together, to set out the final point to which they come. And it has been said that the causes, sources, and elements of independent things are being looked for. But while some independent things are acknowledged by everyone, some people make declarations about some that are peculiar to them; the acknowledged ones are the natural ones, such as fire, earth, water, air, and the other simple bodies, and then the plants and their parts, and the animals, and the parts of animals, and the whole cosmos and the parts of the cosmos, but some people mention peculiarly the forms and mathematical things. But in one way it follows from the discussions that what it is for something to be, and what underlies something, are kinds of thinghood, and in another way that thinghood is the general class, more than the specific one, and the universal more than the particulars; and the forms are also connected with universal and the general class (since it is by the same argument that they seem to be independent things). And since what it is for something to be is thinghood, and the articulation of that is a definition, for that reason distinctions were made about definition and about what something is in virtue of itself; and since a definition is a statement, and a statement has parts, it was also necessary to know about parts—which sort are parts of an independent thing and which not, and if these are the same ones that are parts of the definition. And further, in the course of this, it turned out that neither the universal nor the general class is thinghood, but as for the forms and the mathematical things, one must examine these later, since some people say that these have being apart from perceptible independent things.

But now let us go over what concerns the acknowledged independent things. And these are the perceptible ones. And all perceptible independent things have material. And what underlies something is its thinghood, and in one sense this is the material (and by material I mean that which, while not being actively a *this*, is a *this* potentially),

---

[1] This title for Book VIII supplied by the translator.

1042a 30

but in another sense what underlies something is its articulation and form, which, being a *this,* is separate in articulation; and a third sort of underlying thing is what is composed of these, of which alone there is coming into being and destruction, and which is separate simply. For of the independent things in the sense that corresponds to the articulation, some are separate simply, while others are not.[2]

1042b

But it is clear that the material too is thinghood, for in all changes between contraries, there is something that underlies the changes; for example, in change of place there is something that is now here and in turn elsewhere, and in change by increase there is something that is now just this size and in turn less or more, and in alteration there is something that is now healthy and in turn sick. Similarly, in change of thinghood, there is something that is now in the course of coming into being and in turn in the course of passing away, and is now an underlying thing by being a *this* and in turn an underlying thing by being a lack. And the other kinds of change accompany this one, but it does not accompany one or two of the others, since it is not necessary, if something has material for changing place, that it also have material for becoming and perishing (and what the difference is between becoming simply, and in the sense other than simply, has been stated in the writings on nature[3]).

1042b 10

Chapter 2 Now since thinghood, in the sense of what underlies a thing as its material, is acknowledged, and is thinghood in potency, it remains to say what the thinghood of perceptible things is in the sense of being-at-work. And Democritus seems to think that there are three ways things differ (for he thinks that the underlying body, the material, is one and the same, while what differ are design, which is

---

[2] In the crudest sense, form seems to be superficial and variable, while material underlies it and persists. But the composite whole of material and form also persists through change, and underlies the attributes that come and go. But in the case of animals and plants, it is material that comes and goes, while only form remains intact and underlies each one. In the cosmos, too, material is exchanged and rearranged, while the form stays as it is. In discussing forms, as we are doing now, we are separating them in thought, as universals, but as they are in themselves, as causes of those things that are most properly said to be, their separateness is in no way dependent on our thinking. The inquiry is now in quest of forms as they are in their own right.

[3] See *Physics* 225a 12–20. Any change can be thought of as becoming-something-in-particular, but simple becoming or simple perishing may not belong to the stars, even if they change place. See also 1044b 5–8.

shape, twist, which is position, and grouping, which is order). But it is
obvious that there are many differences; for instance, some things are
spoken of by reference to the composition of their material, as are all
those made by mixing, such as milk blended with honey, others by way
of a binding-cord, such as a bundle, others by means of glue, such as a
book, others by means of bolts, such as a box, others by more than one
of these, others by position, such as a threshold and a capstone (since
these differ by being placed in a certain way), others by time, such as   1042b 20
dinner and breakfast, others by place, such as the winds, and others
by the attributes of perceptible things such as hardness and softness,
density and rarity, or dryness and fluidity, some things differing by
way of some of these, some by all of them, and in all cases, some by
exceeding and others by being exceeded. Therefore it is clear that "is"
is meant in just as many ways; for something is a threshold because it
is placed in a certain way, and its being means its being placed in that
certain way, and being ice means being condensed in a certain way.
And the being of some things will even be defined by means of all of
these, by something being mixed, something else blended, something   1042b 30
else tied together, something else packed together, and something else
using the other differences, such as a hand or a foot.

So one must grasp the kinds of differences (since these will be the
sources of being), such as those by more and less or dense and rare or
the other things of that kind, for all these are ways of exceeding and
falling short. And if anything differs by shape, or by smoothness and
roughness, all these are by means of the straight and the curved. But for
other things, their being will be their being mixed, and their not-being   1043a
the opposite. So it is clear from these considerations that if thinghood
is the cause of each thing's being, it is among these differences that one
must look for what is responsible for the being of each of these things.
Now none of these examples is an independent thing,[4] not even when
linked with something else, but still there is an analogous structure in
each of them; and just as, among independent things, what is attributed
to the material is itself its being-at-work, so too in the other definitions
it is what is most nearly a being-at-work. For example, if one needs to
define a threshold, we will say it is a wooden plank or a stone placed in
such-and-such a way, or of a house that it is bricks and lumber placed

---

[4] Every example given here is an artificial product, or part of a natural one.

1043a 10

thus-and-so (and further, in some cases, that for the sake of which it is), or if it is ice that needs to be defined, that it is water solidified and packed together in such-and-such a way; and harmony is such-and-such a mixing of high and low tones, and one would define in the same way in other cases. Now it is clear from these examples that the being-at-work and the articulation are different for different materials; for of some it is the composition, of others the mixture, and of others some other thing that has been mentioned. That is why, of those who give definitions, some of them, saying what a house is, that it is stones, bricks, and lumber, describe the house in potency, since these things are material, but others, saying that it is a sheltering enclosure for possessions and living bodies, or adding something else of that kind, describe its being-at-work; still others, putting together both of these, describe the third sort of thinghood that is made out of these (for the articulation by means of the differences seems to be a statement of

1043a 20

the form and the being-at-work, while the one that proceeds from the constituents seems rather to be a statement of the material). And it is such definitions that Archytas used to give in a similar way, since they are of both together. For example: What is windlessness? Stillness in an expanse of air. What is a calm? Levelness of the sea; the sea is what underlies it as material, and the being-at-work and form is the levelness.

So from what has been said, it is clear what a perceptible independent thing is, and also in what manner it has its being; for in one way it has being as material, and in another way as form and being-at-work, while in a third sense it is what is composed of these.

1043a 30

Chapter 3   One must not ignore the fact that it sometimes escapes notice whether a name indicates a composite independent thing or its being-at-work and form, for instance whether "house" is a sign jointly for a shelter made of bricks and stones placed in such-and-such a way, or for the being-at-work and form, namely a shelter, or whether "line" indicates twoness in a length or twoness,[5] or whether "animal" means a soul in a body or a soul, since this is the thinghood and being-at-work of a certain kind of body. But "animal" might also be applied to both, not as meaning one articulation, but as pointing to one thing. But

---

[5] See 1036b 12–18 and footnote.

while these things make a difference in some other respects, they make no difference to the inquiry after perceptible thinghood, since what it is for something to be belongs to the form and the being-at-work. For a soul and being-a-soul are the same thing, but being-human and a human being are not the same, unless the soul is going to be called a human being, and in that way they are the same in a certain way, but in a certain way not.

    Now it is obvious to those who inquire about it that a syllable is not made of its letters plus combination, nor a house out of bricks plus combination, and rightly so, for neither combination nor mixture is among those things of which they are the combination or the mixture. And similarly too with all other things; for example, if a threshold is what it is by placement, then the placement does not come from the threshold but rather the latter from the former.[6] Nor indeed is humanness animalness plus two-footedness, but there has to be something which is apart from these, since these are its material, and that something is neither an element nor derived from an element, but since they leave this out, people describe its material. So if this is responsible for the being and thinghood of a thing, one could call this the thinghood itself. (Now it is necessary that this either be everlasting, or else be destructible without being in the process of being destroyed and have come into being without being in the process of becoming.[7] But it has been demonstrated and made clear in other chapters that no one makes or generates the form, but a *this* is made and something composed of form and material comes into being. But whether those things that are the thinghood of destructible things have being separately is not at all clear yet, except that it is clear for certain things at least that this is not possible, as many as are not capable of being apart from the particulars, such as a house or piece of furniture. So presumably these things themselves are not independent things, nor is any of the other things that are not composed by nature, for one may posit that nature alone is the thinghood in destructible things.[8])

1043b

1043b 10

1043b 20

---

[6] That is, the placement does not come out of the threshold as one of its constituents, but the threshold comes out of the placement as effect from cause.

[7] If this extra something could undergo destruction or generation, it would itself be a composite, and would need something else to make it whole.

[8] A house is relatively stable, and so resembles a genuine independent thing, but nothing is at work in it maintaining it as a house. Human art sets the natural tendencies

And so there is a certain appropriateness to the impasse that the Antistheneans and other crude people of that sort[9] used to raise, claiming that it is impossible to define what anything is (since the definition is just beating around the bush), though it is possible to teach what sort of thing something is, as with silver, one cannot say what it is, but that it is like tin; therefore it does belong to an independent thing to be capable of a definition and an account, namely to a compound 1043b 30   one, whether it is intelligible or perceptible, but no longer to those things of which this first consists, so long as a defining statement indicates something attributed to something and requires that there be something as material and something else as form.

And it is also clear for this reason that, if independent things are in some way numbers, it is in this way that they are composed of units and not in the way that some people say, for a definition is a certain kind of number, since it too is divisible into indivisible parts (since articulations are not infinite), and that is the sort of thing a number is. And just as, if something is subtracted from or added to the units of which a number is composed, it is no longer the same number but a 1044a   different one, even if the tiniest bit were subtracted or added, so too, neither the definition nor what it is for something to be will any longer be what it is if anything is subtracted or added. And it is necessary to a number that there be something by means of which it is one, but as things stand, the people who speak of thinghood as numbers are not able to say by means of what each is one (for either each is not one but is like a heap, or if it is one, it needs to be said what it is that makes one thing out of many); and a definition is one, but similarly they cannot account for this. And it turns out this way reasonably, for it belongs to the same argument that thinghood is also one in the same way, not however in the way they say it is, as though it were a unit or a point, but each independent thing is a complete being-at-work-staying-itself,

---

of the materials in it against each other, so that the roof, in trying to fall, holds up the walls, while the walls, in trying to fall, hold up the roof. But any homeowner knows that the only inherent activity of the house as such is to fall apart.

[9] Antisthenes was the first of the people called Cynics. They were not philosophers but public moralists, attacking conventional beliefs and scorning wealth. In Plato's *Theaetetus*, beginning at 201E, Socrates gives a more philosophic version of this argument (his "dream," since in fact the Antistheneans came along after his death). Aristotle's example of the syllable, and his comparison to a number, as instances of wholeness and unity, come from Socrates's refutation of the argument.

and a particular nature. And just as a number does not have any more   1044a 10
or less, neither does thinghood in the sense of form, but if at all, only
thinghood that includes material.

So about the coming into being and destruction of what are called
independent things,[10] in what sense they admit of it and in what
sense they are incapable of it, and about the tracing back of things
to numbers, let distinctions have been made to this extent.

Chapter 4   Concerning the thinghood of material things, one must
not overlook the fact that, even if all things are made out of the same
first constituent or the same primary elements, and the same material
is the source of their coming-into-being, still there is some material
peculiar to each kind of thing, as of phlegm, what is sweet or fatty, or
of bile, what is bitter or some other things, though perhaps these are   1044a 20
made of the same thing. And there come to be a number of materials
for the same thing whenever one of them is material for another, as
is the case with the phlegm composed of what is fatty and sweet if
what is fatty is made of what is sweet, or of bile, by a decomposition
into the first bile material. For one thing comes from another in two
senses, either because it will be further down the road or because it
arises when something is decomposed back into its source. And it is
possible, when the material is one, for different things to come from it
on account of the cause that sets it in motion, as out of wood either a
box or a bed. But in some cases the material has to be different if the
things are different; for instance, a saw could not be made out of wood,
nor does this depend on the cause that sets its coming-into-being in
motion, since it will not make a saw out of wool, or out of wood either.
If, therefore, the same thing admits of being made out of a different   1044a 30
material, it is clear that the art and origin that sets its coming-into-
being in motion is the same, for if both the material and the mover

---

10 At 1043b 14, Aristotle has made the crucial move out of the perceptible realm,
saying that thinghood is most properly ascribed to whatever it is that *causes* a thing
to be one and whole, the something else that is not an element or constituent of it.
In Bk. VII, Ch. 16, he concluded that most of the so-called independent things are
only parts and potencies, that is dependent things; now he has concluded that even
the animals, plants, and cosmos, each of which is a complete being-at-work-staying-
itself, are dependent on that something else that causes them, which is no part of the
world of sensory experience, but has been deduced as underlying it and responsible
for it.

were different, the thing produced would be too. So whenever one is inquiring after what is responsible for something, since causes are meant in more than one way, one must state all the causes the thing admits of. For example, what is the cause of a human being in the sense of material? Isn't it the menstrual fluid? And what is the cause that sets the coming-into-being in motion? Isn't it the semen? And what is the cause in the sense of form? What it keeps on being in order to be. And what is the cause for the sake of which it is? Its end, though presumably both of the last two causes are the same. And one must state the nearest causes: What is the material? Not fire or earth but the material peculiar to the thing.

1044b

So about the natural independent things that come into being, it is necessary to approach them in this way if one is going to approach them in the right way, if in turn the kinds of cause are these and this many and one has to know the causes, but in the case of the independent things that are natural but everlasting, it is another story. For perhaps some of them do not have material, or not this sort but only material for change of place. Nor is there material for those things that are by nature, but are not independent things, though what underlies them is an independent thing. For instance, what is the cause of an eclipse— what material? There is none, but it is the moon that undergoes it. But what is responsible for it in the sense of setting it in motion and wasting away the light? The earth. And presumably there is nothing it is for the sake of.[11] The formal cause is the articulation of it, but the articulation is unrevealing if it does not include the cause. For example, what is an eclipse? A deprivation of light. But if one adds "by the action of the earth's coming to be in the middle," this is the articulation with the cause in it. But with sleep, it is not clear what is the primary thing that undergoes it. But isn't it the animal? Yes, but this in which part, which one primarily? The heart, or something else. And then, by the action of what? And then, what is it that is undergone, that happens to that

1044b 10

---

[11] This goes together with the fact that the eclipse does not belong to any independent thing, and does not contribute to its wholeness. It is a relation among the earth, moon, and sun, and incidental to them all. Aristotle likes the example of the eclipse (which he repeatedly uses in the *Posterior Analytics*) because the earth literally mimics the middle term of a syllogism, causing a connection between the extreme terms (moon and sun).

part and not to the whole? Isn't it a certain kind of motionlessness? Yes, but this on account of the primary part's undergoing what?   1044b 20

Chapter 5   But since some things are or are not without coming into being or being destroyed, such as points, if they are at all,[12] or in general the forms (for whiteness does not come into being, but wood becomes white, if everything that comes into being becomes something from something), not all contraries could arise out of one another, but a pale person comes from a swarthy one in a different way from that in which white comes from black; and there is not material in everything, but in those things of which there is a coming-into-being and a change into one another. So there will not be material in those things that are or are not without changing.

There is an impasse about how the material of each thing is related   1044b 30
to contraries. For example, if the body is potentially healthy, and the contrary of health is sickness, is it potentially both? And is water potentially wine and vinegar? Or is it material for the former as a result of an active condition and a form, but for the latter as a result of a deprivation and decay that is contrary to nature? But there is also an impasse about why wine is not material for vinegar nor potentially vinegar (even though vinegar comes from it), and why a living thing is not potentially a corpse. Or is it not so, but rather the decays are   1045a
incidental to the animal and the wine, while the material of the animal is itself, by virtue of decay, a potency and material for a corpse, and water for vinegar, since the decayed things come from the others only in the sense that night comes from day. And all the things that change into one another in that way have to turn back into their material, such as if an animal is to come from a corpse, it must turn first into its material, and in that way next into an animal, or vinegar into water, and in that way then into wine.

Chapter 6   And in regard to the impasse stated about both definitions and numbers, what is responsible for their being one? For all things that have more than one part, and of which the sum is not like a heap, but a whole that is something over and above the parts, have   1045a 10

---

[12] Euclid says "let there be a point A," and there is one. Aristotle regards them only as limits of lines. See above, 1002a 33–1002b 7.

something that is responsible for them; since among bodies, the cause of the being-one of some of them is contact, and of others stickiness or some other attribute of that sort. But a definition is one statement not by being bundled together like the *Iliad,* but by being *of* one thing.[13] What then is it that makes humanness one, and why is it one and not many, such as animal plus two footed, both otherwise but especially if, as some people say, there is some animal-itself and two-footed-itself? For why is humanness not those things themselves, and why will humans not be what they are by partaking not of humanness nor of one thing but of two, animalness and two-footedness, so that humanness would not be one thing at all but more than one, namely animal plus two footed?

1045a 20

Now it is clear that, for those who approach defining and explaining in this way that they are accustomed to, it is not possible to give an account of it and resolve the impasse. But if, as we say, there is one thing that is material and one that is form, and the former has being as potency and the latter as being-at-work, the thing sought after would no longer seem to be an impasse. For this impasse is the same one there would be if the definition of "overcoat" were a round bronze thing; for the name would be a sign of this formulation, so that the thing sought after is what is responsible for the being-one of the roundness and the bronze. But no impasse any longer presents itself, because the one is material and the other is form. What, then, is responsible for this, for the active being of what was in potency, other than the maker, in those things of which there is a coming-into-being? For no other thing is responsible for the potential sphere's being actively a sphere, but that is what it is for it to be in either case.[14] But there is one kind of material that is intelligible and another kind that is perceptible, and one part of the statement is always material and the other part being-at-work, as a circle is a figure with respect to a plane. But as many things as do

1045a 30

---

[13] We are presumably to imagine the twenty four papyrus rolls of the *Iliad* tied up with string, since in the *Poetics* (1451a), Aristotle says the poem is built around one action.

[14] The sphere in potency is the composite thing, looked at from the standpoint of the bronze; the sphere in activity is the same thing, looked at from the standpoint of its roundness. In the case of humanness (substituting rational for two footed), its intelligible material is animality looked at as already pervaded by and straining toward rationality, while its being-at-work is the activity combined with reason that inheres in animality, and forms it into humanness.

not have either intelligible or perceptible material, are each of them
immediately some very thing that is one, just as also some very thing   1045b
that *is,* a *this,* an of-this-sort, a so-much (and this is why neither being
nor one is included in definitions), and what it is for something to be
is immediately a particular *one* and a particular being. Hence there is
no other thing responsible for the being-one of any of these, nor of the
being-a-being of each, since each is immediately a certain being and a
certain *one,* not in the sense of being in a class of beings or ones, nor of
being among things that have being apart from particulars.

On account of this impasse, some people talk about participation,
and are at a loss about what is responsible for participation and what
participating is; others talk about co-presence, just as Lycophron says   1045b 10
that knowledge is a co-presence of knowing and a soul, while still
others say life is a composition or conjunction of a soul with a body. And
yet the same formulation applies to everything: for being healthy will
be a co-presence or conjunction or composition of a soul and health,
and the bronze's being a triangle will be a composition of bronze and
triangle, and white will be a composition of a surface and whiteness.
And the reason they say these things is that they are looking for a
formulation that unites potency and complete being-at-work, plus a
difference. But as was said, the highest level of material and the form
are one and the same thing,[15] the former potentially, the latter actively,
so that looking for what is responsible for their being one is like looking   1045b 20
for a cause of one thing; for each of them is a certain *one,* and what is in
potency and what is in activity are in a certain way one thing. Therefore
there is nothing else responsible, unless in the case of something that
moves it from potency to being-at-work, but every thing that does not
have material is simply something that is itself one.

---

[15] The strongest statement of this is in *On the Soul,* at 412b 4–9, where Aristotle says
that there is no more question about whether a soul and body, or any thing and its
material, are one, than there is whether the wax and the shape pressed into it are
one, since complete being-at-work-staying-itself is exactly what *one* and *being* mean
in the sense that is primary and governing for each. So the conclusions of Book VIII
are (a) form is not an arrangement of parts but a being-at-work of material that has
the potency for it, (b) the thinghood of a thing is its form, and (c) the form itself has
an internal form/material structure, and is therefore unified as intelligible material-at-
work. The whole relation of potency and being-at-work must now be examined, as
the next step in uncovering the causes and sources of being as being.

Book IX (Book Θ)
Potency and Being-at-Work[1]

Chapter 1 What concerns being of the primary sort, toward which 1045b 27
all the other ways of attributing being are traced back, has been
discussed, namely what concerns thinghood (for the other sorts of
being are articulated in virtue of the articulation of the thinghood of a
thing, the how-much, the of-what-sort, and the other things attributed 1045b 30
in that way, since all of them will include the articulation of thinghood,
as we say in the first chapters); but since being is spoken of in one way
by way of what or of what sort or how much something is, but in
another way in virtue of potency and complete being-at-work, and of
a doing-something, let us make distinctions also about potency and
complete-being-at-work, and first about potency in the sense in which
it is meant most properly, although it is not the sense that is most useful
for what we now want. For potency and being-at-work apply to more 1046a
than just things spoken of in reference to motion. But when we have
discussed them in this sense, we will make clear their other senses in
the distinctions that concern being-at-work.

Now it has been distinguished by us in other places that potency
and being potential are meant in more than one way, and of these, let
the ones that are called powers ambiguously be set aside (for some
things are so called by means of some likeness, as in geometry we
speak of powers and incapacities by reference to something's being or
not being a certain way[2]), but as many of them as point to the same
form are all certain kinds of sources, and are meant in reference to one 1046a 10
primary kind of potency, which is a source of change in some other
thing or in the same thing as other.[3] For one kind is a power of being
acted upon, which is a source in the acted-upon thing itself of passive
change by the action of something else or of itself as other; another is

---

[1] This title for Book IX supplied by the translator.

[2] See 1019b 33–34 and note.

[3] This constant qualification, used here and in Bk. V, Ch. 12, serves to distinguish
potencies from a thing's nature. A doctor doctors himself as another, as a patient who
happens to be himself, but a cut in his finger heals itself in its own right; medical skill
is a potency, while the self-maintenance of a being as a whole is a nature. See 1049b
9–11.

an active condition of being *un*affected for the worse or for destruction by the action of a source of change, either some other thing or itself as other. For the articulation of the primary sort of potency is present in the definitions of all of these.[4] And these potencies in turn are spoken of either as only acting or being acted upon, or as acting or being acted upon well, so that in the descriptions of these potencies too, there are present in a certain way the descriptions of the kinds of potency that are more primary than they are.

1046a 20   And it is clear that there is a sense in which the potency of acting and being acted upon is one (since something is potential both by means of its own potency to be acted upon and by something else's potency to be acted upon by it), but there is a sense in which they are different. For the one is in the thing acted upon (for it is by virtue of having a certain kind of source, and even because its material is a certain kind of source, that the acted-upon thing is acted upon, even though it is one thing and is acted upon by the action of another—for something oily is burnable and something that gives way just so is crushable, and similarly in other cases), but the other is in the thing acting, as heat and the housebuilding power are in, respectively, something that confers heat and someone who can build houses; hence, insofar as something has developed as a natural whole, it cannot be passive to itself, since it is one thing and there is no other. And lack of capacity, or something

1046a 30   incapable, is a deprivation opposite to this sort of potency, so that every potency is contrary to an incapacity in the same thing, for the same thing. But deprivation is meant in more than one way; for it is the not-having what something would naturally have, either at all, or when it would naturally have it, and either in the way that it would naturally have it, such as completely, or just in any respect whatever. And in some cases, when things that naturally have something do not have it on account of force, we speak of them as lacking it.

---

[4] The primary sense of potency is "source of change in another or as another." The sense of mere logical possibility is one of the ambiguous capacities dismissed above. Even to be acted upon, a thing must contain its own source of being acted upon, since it could also have an active condition of imperviousness. Virgil's Aeneas is described as resisting his own compassion for Dido as an oak tree resists the wind (*Aeneid* Book IV, lines 441–449), but Dante describes his acceptance of Beatrice's pity as the uprooting of an oak (*Purgatory* Book XXXI, lines 70–73). Both conditions originate in the people who have them, and both require work.

Chapter 2 But since some sources of this kind are present in things without souls, and others in things with souls, both in the soul in general and in the part of it that has reason, it is obvious that of potencies too, some will be irrational and some will include reason; and this is why all the arts and the productive kinds of knowledge are potencies, since they are sources of change in another thing, or in the same thing as other. And all potencies that include reason are themselves capable of contrary effects, but with the irrational ones, one potency is for one effect, as something hot has a potency only for heating, while the medical art is capable of causing disease or health. And the reason is that knowledge is a reasoned account, while the same reasoned account reveals both a thing and its lack, though not in the same way, and in a sense pertains to both, though more so to its proper subject, so that such kinds of knowledge are necessarily about contrary things, though about the one sort in their own right and the other sort not in their own right; for the account is about the one in virtue of itself, but about the other in a certain incidental way, since it reveals the contrary by means of negation and removal, since the thing that something primarily lacks is its contrary, and this is the removal of that other thing.

Now since contraries do not come to be present in the same thing, while knowledge is a potency that has reason, and the soul is a source of motion, then even though something healthful produces only health and something that heats produces only heat and something that cools produces only cold, the person who knows produces both contraries. For a reasoned account pertains to both, though not in similar ways, and is in a soul that has a source of motion, so that it will set both contraries in motion from the same source, connecting them to the same account; which is why things that are potential in virtue of reason act in ways contrary to things that are potential without reason, since contrary things are contained in one source, the reasoned account. And it is clear that, with the potency of doing something well, the potency of merely doing or suffering it follows along, while the former does not always follow along with the latter, since the one doing something well necessarily also does it, but the one merely doing it does not necessarily also do it well.[5]

---

[5] The last sentence is a reminder that the positive potency is always primary. The

Chapter 3   There are some people, such as the Megarians, who say

that something is potential only when it is active, but when it is not active it is not potential; for instance someone who is not building a house is not capable of building a house, but only the one who is building a house, when he is building a house, is capable of it, and similarly in other cases. The absurd consequences of this opinion are not difficult to see. For it is clear that someone will not even be a house builder if he is not building a house (since to be a house-builder is to be capable of building a house), and similarly too with the other arts. So if it is impossible to have such an art if one has not at some time

learned and taken hold of it, and then impossible not to have it if one has not at some time lost it (and this is either by forgetfulness or by some affliction or by time, for it is not by the destruction of the thing with which the art is concerned, since it is always present), whenever one stops he will not have the art, so if he starts building a house again straight off, how will he have acquired the art? And it is similar with things without souls; for there will not be anything cold or hot or sweet or perceptible in any way when it is not being perceived, so that these people turn out to be stating the Protagorean claim.[6] But nothing will have perception either, if it is not perceiving and doing so actively. So if what does not have sight, and is of such a nature as to have it, when it is of such a nature and moreover has being, is blind, then the same

people will be blind many times during the same day, and deaf.

What's more, if what is lacking a potency is incapable, what is not happening will be incapable of happening; but of what is incapable of happening, it is false for anyone to say either that it is so or that it will be so (since that is what incapable means), so that these assertions abolish both motion and becoming. For what is standing will always be standing and what is sitting always sitting, since it will not stand up if it is sitting, since what does not have the potency of standing up will be incapable of standing up. So if these things do not admit

---

structure is not of a neutral capacity with something overlaying it that directs it one way or the other, but of a positive capacity that inherently goes together with a certain knowledge, but can incidentally be turned in the negative direction. The survey of kinds of potency in Ch. 1 and 2 repeatedly finds the same pattern: even passive potencies are sources, even the potency to resist change is an active condition, and even a two-sided potency is primarily directed to something's good.

[6] See note to 1007b 22. Protagoras and this claim of his play a central role in Plato's *Theaetetus,* which is presented as narrated by one Megarian to another.

of being said, it is clear that potency and being-at-work are different (for these assertions make potency and being-at-work the same, for which reason it is no small thing they are seeking to abolish), so that it is possible for something to be capable of being yet not be, or be capable of not being and yet be, and similarly with the other ways of attributing being, that something that is capable of walking not be walking, or that is capable of not walking be walking. What is capable is that which would be in no way incapable if it so happened that the being-at-work of which it is said to have the potency were present.[7] I mean, for instance, that if something is capable of sitting, and admits of sitting, if it so happens that sitting is present to it, it will in no way be incapable of it, and similarly if something is capable of being moved or of causing motion, or of standing or making something stand, or of being or becoming, or of not being or not becoming.

1047a 20

And the phrase being-at-work, which is designed to converge in meaning with being-at-work-staying-complete, comes to apply to other things from belonging especially to motions, since being-at-work seems to be motion most of all, and this is why people do not grant being-in-motion to things that do not have being, though they do allow them other attributes, for instance that things that do not have being are thinkable or desirable, but are not moved, and this is because, while not actively being, they would have to be in activity. For of the things that are not, some are potentially; but they are not, because they are not at-work-staying-complete.

1047a 30

1047b

Chapter 4   But if what has been said is what is potential, in the sense that it is a necessary accompaniment to it, it is clear that it cannot be true to say that such-and-such is possible, but will not be the case (as a result of which statement, things incapable of being the case would disappear); I mean, for instance, the case of someone who would say that the diagonal of a square is capable of being measured, even though it will not be measured—someone, that is, who does not reckon there to be anything impossible—because nothing prevents anything from being capable of being or becoming so, that is not or is not about to be

---

[7] Scholars worry about whether this is meant to be a definition of potency, but that has been given already as "a source of change in something else, or in the same thing as other." The point of this sentence is to emphasize the conditional relation such a source has to that of which it is the source.

1047b 10   so. And that is a necessary consequence of the things set down, if we also assume that, if what is not the case but is possible were to be or come about, there would be nothing impossible; but this will turn out to be impossible, since the diagonal is incapable of being measured. For the false and the impossible are not the same thing; for that you are now standing is false, but not impossible.

At the same time, it is also clear that, if it is necessary for B to be the case when A is, it is also necessary for B to be capable of being the case when A is capable of being the case; for if it is not necessary that B be capable of being so, nothing prevents it from being incapable of being so. Now let A be possible. Accordingly, at the time when A would be capable of being the case, if A were established, nothing incapable of being the case would result; but B would then necessarily be the case.
1047b 20   But it was taken to be impossible. So let it be impossible. But now if B is incapable of being the case, A is necessarily also impossible. But the first one was taken to be impossible, and therefore the second is too. Therefore, if A is to be possible, B will also be possible, so long as they are so related that when A is the case, B necessarily is too. So if, when A and B are so related, B is not possible, in that case, A and B will not have the relation that was assumed; and if, when A is possible, B must necessarily be possible, then if A is the case, B must necessarily also be the case. For B's necessarily being capable of being the case when A is possible means, if A is the case, when and in the way that it was
1047b 30   capable of being the case, B also has to be the case then and in that way.

Chapter 5   Of all potencies, since some are innate, such as the senses, while others come by habit, such as that of flute playing, and others by learning, such as that of the arts, some, those that are by habit and reasoning, need to have previous activity, while the others that are not of that kind, and apply to being acted upon, do not need it. And
1048a   since what is potential is capable of something, at some time, in some way, and all the other things that need to be added in its delineation, and some things are capable of causing motion in accordance with reason and their potencies include reason, while other things are unreasoning and their potencies are irrational, the former must be in things with souls, but the latter in both kinds of beings; with potencies of the latter sort, it is necessary, whenever a thing that is active and one that is passive in the sense in which they are potential come near each

other, that the one act and the other be acted upon, but with the former
sort this is not necessary. For with all these unreasoning potencies, one
is productive of one effect, but those others are productive of contrary
effects, so that each would at the same time do contrary things; but
that is impossible. It is necessary, therefore, that there be something   1048a 10
else that is governing; by this I mean desire or choice. For whatever
something chiefly desires is what it will do whenever what it is capable
of is present and it approaches its passive object; and so everything
that has a potency in accordance with reason must do this whenever
it desires that of which it has the potency and in the way that it has
it, and it has it when the passive object is present and is in a certain
condition. If this is not so, it will not be capable of acting (for it is not
necessary to add to the description "when nothing outside prevents"
since it has the potency in the sense that it is a potency of acting, and
this is not in every situation but when things are in certain conditions,
among which conditions the outside obstacles will be eliminated, since   1048a 20
some of the things in the description remove them); and so, even if one
wishes and desires to do two things at the same time or to do contrary
things, one will not do them, for one does not have the potency for
them in that way, and there is no potency for doing them at the same
time, since a thing will do the things it is capable of in the way it is
capable of them.

Chapter 6   Since what concerns the kind of potency that corresponds to motion has been discussed, let us make distinctions about
being-at-work, to mark out both what it is and what sort of thing it is.
For that which is potential will also be clear at the same time to those
who make these distinctions, since we speak of the potential not only
as that which is of such a nature as to move some other thing or be
moved by something else, either simply or in a certain respect, but   1048a 30
also in another way, and it is because we are inquiring after that other
meaning that we went through this one. Now being-at-work is something's being-present not in the way that we speak of as in potency;
and we speak of as being in potency, for example, Hermes in a block
of wood or a half line in the whole, because they can be separated
out, or someone who knows, even when he is not contemplating, if
he is capable of contemplating. The other way these things are present
is in activity. And what we mean to say is clear by looking directly
at particular examples, nor is it necessary to look for a definition of

1048b

everything, but one can also see at one glance, by means of analogy, that which is as the one building is to the one who can build, and the awake to the asleep, and the one seeing to the one whose eyes are shut but who has sight, and what has been formed out of material to the material, and what is perfected to what is incomplete. And let being-at-work be determined by one part of each distinction, and what is potential by the other. But not all things that are said to be in activity are alike, other than by analogy: as this is in respect to or in relation to that, so is this thing here in respect to or in relation to that thing there. For some of them are related in the manner of a motion to a potency, others in the manner of thinghood to some material.[8]

1048b 10

But it is in a different way that the infinite and the void, and anything else of that kind, are said to be in potency or in activity, as opposed to most of the things that have being, such as whatever sees or walks or is seen. For the latter admit of simply being true at some time (for in one sense something is a thing seen because it is being seen, in another sense because it is capable of being seen), but the infinite is not potential in the sense that it is going to be actively separate, except in knowledge. For the fact that an infinite division does not come to an end permits this sort of being-at-work to be in potency, though not to be separate.

1048b 20

And since, of the actions that do have limits, none of them is itself an end, but it is among things that approach an end (such as losing weight, for the thing that is losing weight, when it is doing so, is in motion in that way, although that for the sake of which the motion takes place is not present), this is not an action, or at any rate not a complete one; but that in which the end is present is an action. For instance, one sees and is at the same time in a state of having seen, understands and is at the same time in a state of having understood, or thinks contemplatively and is at the same time in a state of having thought contemplatively, but one does not learn while one is at the same time in a state of having learned, or get well while in a state of having gotten well. One does

---

[8] This is the reason being-at-work is not definable. Potency is clearest to us in the capacity for motion or change, but motion itself is defined as a form of being-at-work-staying-complete. But being-at-work is usually reserved for the activities that are not motions. In Aristotle's discourse, *energeia* is an ultimate explanatory term, not itself explainable by anything simpler or clearer, but Aristotle has chosen the word to carry meaning on its own, and translating it as "actuality" kills it.

live well at the same time one is in a state of having lived well, and one
is happy at the same time one is in a state of having been happy. If this
were not so, the action would have to stop at some time, just as when
one is losing weight, but as things are it does not stop, but one is living
and is in a state of having lived. And it is appropriate to call the one
sort of action motion, and the other being-at-work. For every motion is
incomplete: losing weight, learning, walking, house-building. These
are motions, and are certainly incomplete. For one is not walking and
at the same time in a state of having walked, nor building a house and
at the same time in a state of having built a house, nor becoming and
in a state of having become, nor moving and in a state of having been
moved, but the two are different; but one has seen and at the same time
is seeing the same thing, and is contemplating and has contemplated
the same thing. And I call this sort of action a being-at-work, and that
sort a motion. So that which *is* by way of being-at-work, both what it is
and of what sort, let it be evident to us from these examples and those
of this kind.

Chapter 7   Now when each thing is in potency and when not must
be distinguished, since it is not the case at just any time whatever. For
example, is earth potentially a human being? Or is it not, but rather is
so only when it has already become germinal fluid, and perhaps not
even then? Then it would be just as not everything can be healed, by
either medical skill or chance, but there is something that is potential,
and this is what is healthy in potency. But the mark of what comes to
be in complete activity out of what has being in potency, as a result of
thinking, is that, whenever it is desired it comes about when nothing
outside prevents it, and there, in the thing healed, whenever nothing
in it prevents it. And it is similar too with a potential house; if nothing
in this or in the material for becoming a house stands in the way,
and there is nothing that needs to have been added or taken away or
changed, this is potentially a house, and it is just the same with all
other things of which the source of coming into being is external. And
of all those things in which coming into being is by means of something
they have in themselves, those are in potency which will be on their
own if nothing outside blocks their way; for instance, the semen is not
yet potential (since it has to be in something else, and to change), but
whenever, by virtue of its own source of motion, it is already such,
in this condition it is from that point on in potency, though in that

1048b 30

1049a

1049a 10

previous condition it has need of another source, just like earth that is not yet potentially a statue (since once it changes it will be bronze).

1049a 20   And it seems that what we speak of as being not this but made of this—for example a box is not wood but wooden, and the wood is not earth but earthy, and the earth in turn, if it is not some other thing but just made-of-that—the thing next following is always simply potentially the one that precedes it. For instance, the box is not made of earth, nor is it earth, but is wooden, since this is potentially a box, and the material for a box is this, simply so for a box considered simply, or this particular wood for this particular box. But if something is the first thing that is no longer said to be made-of-this in reference to any other thing, this will be the first material; for instance, if earth is airy, while air is not fire but fiery, then fire is the first material, if it is not a *this*. For those things to which something is attributed, that is, underlying things, differ in this way: by being or not being a *this*. For example, a human being is something that underlies its attributes,[9]

1049a 30   and is both a body and a soul, and the attribute is cultivated or pale (and when cultivation has come to be in it, it is said to be not cultivation but cultivated, and a human being is not paleness but pale, and is not motion or the act of walking but is something that moves or walks, like the thing that is said to be made-of-that), and so the last of the things that underlie something in this way is an independent thing. But the last of those things that do not underlie in that way, but to which the thing attributed is some form or a *this,* is material and its thinghood is

1049b   of a material sort. And it turns out rightly that of-that-sort applies to material and to attributes, since both are indeterminate.

So when something should be spoken of as potential and when not have been said.[10]

---

[9] To underlie is a synonym for having something attributed. It is not these two notions that are being distinguished, but two ways that a thing can underlie what is attributed to it.

[10] It is now fully explicit that Aristotle does not use the word potential to speak of everything that is possible. It is the immediately neighboring material, ready to pass into something when nothing prevents it, if it is natural, or when the artisan wishes and nothing prevents it, if it is artificial, that has a potency. Aristotle takes seriously the testimony of sculptors that Hermes is in the stone or the wood (1048a 34–36 and 1017b 1–8); that is, the material even of art is not infinitely moldable by an artist's creativity, but is a contributing source in its own right to the product that comes of it. In nature, the potency of material is the whole story, so much so that an internal potency is what Aristotle means by nature.

Chapter 8   And since the various ways in which something is said to take precedence have been distinguished,[11] it is clear that being-at-work takes precedence over potency. And I mean that it takes precedence not only over potency as defined, which means a source of change in another thing or in the same thing as other, but over every source of motion or rest in general. For nature too is in the same general class as potency, since it is a source of motion, though not in something else but in a thing itself as itself. Now over everything of that kind, being-at-work takes precedence both in articulation and in thinghood, but in time, there is a sense in which it does and a sense in which it does not. That it takes precedence in articulation is obvious (for it is by admitting of being at work that the primary sort of potency is potential—I mean, say, that the one with the house-building power is the one who is capable of building a house, the one with the power of sight is the one capable of seeing, and the thing with the power of being seen is the thing capable of being seen, and the same description also applies to the rest, so that the statement of the being-at-work must be present beforehand, and the knowledge of it must precede the knowledge of the potency). In time it takes precedence in this way: a thing that is the same in form, though not numerically the same, that is at work, takes precedence. By this I mean that, over this particular human being that is already present in activity, or over this particular grain or this particular thing that is seeing, the material or the seed or the one capable of seeing takes precedence in time, which are potentially a human being or grain or one who is seeing, but are not yet so in activity; but preceding these in time, there are other things that are at work, out of which these particular ones were generated, for what is at work always comes into being from what is in potency, by the action of what is at work, a human being from a human being, say, or someone cultivated by the action of someone cultivated—some mover is always first, and what causes motion is already at work. And it was said in the chapters about thinghood that everything that comes into being becomes something from something, and by the action of something which is the same in form.[12]

1049b 10

1049b 20

---

[11] See Bk. V, Ch. 11, and 1028a 33–35.

[12] For the way that this is true of artifacts as well as of natural beings, see especially 1032b 11–15.

1049b 30   And this is why it seems to be impossible to be a house-builder if one has not built any houses, or a harpist if one has not played the harp at all; for the one learning to play the harp learns *to* play the harp by playing the harp, and similarly with others who learn things. And from this there arises the sophistical objection that someone who does not have knowledge would be doing the things that the knowledge is about, since the one learning does not have knowledge. But since something of what comes into being has always already come into being, and in general something of what is in motion has

1050a   always already been moved (which is made clear in the writings about motion[13]), presumably the one who is learning must also already have something of knowledge. But then it is also clear from the same considerations that being-at-work takes precedence in this way too over potency, in respect to becoming and time.

But surely it takes precedence in thinghood too, first because things that are later in coming into being take precedence in form and in thinghood (as a man does over a boy, or a human being over the germinal fluid, since the one already has the form, and the other does not), and also because everything that comes into being goes up to a source and an end (since that for the sake of which something *is* is a source, and coming into being is for the sake of the end), but the being-at-work is an end, and it is for the enjoyment of this that

1050a 10   the potency is taken on. For it is not in order that they may have the power of sight that animals see, but they have sight in order to see, and similarly too, people have the house-building power in order that they may build houses, and the contemplative power in order that they may contemplate; but they do not contemplate in order that they may have the contemplative power, unless they are practicing, and these people are not contemplating other than in a qualified sense, or else they would have no need to be practicing contemplating.

What's more, material is in potency because it goes toward a form; but whenever it is at work, then it is in that form. And it is similar in

---

[13] In Bk. VI, Ch. 6, of the *Physics,* Aristotle proves that there is no first instant of motion or change. There is a more directly relevant passage in Book VII of that work, 247b 1–248a 9, in which he argues that knowledge is always already active in us, though obscured by distracting disorderly motions. The primary kind of learning is like recollecting, an emerging habit of concentrating on what is already going on in us.

other cases, even those of which motion is itself the end, and that is why
teachers display a student at work, thinking that they are delivering
up the end, and nature acts in a similar way.[14] For if things did not
happen this way, they would be like the Hermes of Pauson, since it   1050a 20
would be unclear whether the knowledge were inside or outside, just
as with that figure. For the end is work, and the work is a being-at-
work, and this is why the phrase being-at-work is meant by reference
to work and extends to being-at-work-staying-complete.[15]

But since the putting to use of some things is ultimate (as seeing
is in the case of sight, in which nothing else apart from this comes
about from the work of sight), but from some things something comes
into being (as a house, as well as the activity of building, comes from
the house-building power), yet still the putting to use is no less an
end in the former sort, and in the latter sort it is more an end than
the potency is; for the activity of building takes place within the thing
that is being built, and it comes into being and *is* at the same time as
the house. So of those things from which there is something else apart   1050a 30
from the putting-to-use that comes into being, the being-at-work is in
the thing that is made (as the activity of building is in the thing built
and the activity of weaving in the thing woven, and similarly with the
rest, and in general motion is in the thing moved); but of those things
which have no other work besides their being-at-work, the being-at-
work of them is present in themselves (as seeing is in the one seeing
and contemplation in the one contemplating, and life is in the soul,
and hence happiness too, since it is a certain sort of life). And so it   1050b
is clear that thinghood and form are being-at-work.[16] So as a result
of this argument it is obvious that being-at-work takes precedence

---

[14] How does nature display that a squirrel has reached the completion for the sake
of which it exists? In the spectacle of the squirrel at work being a squirrel.

[15] That is, beings do not just happen to perform strings of isolated deeds, but
their activity forms a continuous state of being-at-work, in which they achieve
the completion that makes them what they are. Aristotle is arguing that the very
thinghood of a thing is not what might be hidden inside it, but a definite way of
being unceasingly at-work, that makes it a thing at all and the kind of thing it is.

[16] This is the next-to-final conclusion Aristotle reaches about being. The following
sentence is followed to a conclusion in Book XII, where the highest source of being is
uncovered.

over potency in thinghood, and as we have been saying, one being-at-work always comes before another in time until one always reaches the being-at-work of something that first causes motion.

But being-at-work takes precedence in an even more governing way; for everlasting things take precedence in thinghood over destructible ones, and nothing that is in potency is everlasting. And the reason is this: every potency belongs at once to a pair of contradictory things,[17] for what is not capable of being present cannot belong to anything, but everything that is potential admits of not being at work. Therefore, what is capable of being admits both of being and of not being, and so the same thing is capable both of being and of not being. But what is capable of not being admits of not being, and what admits of not being is destructible, either simply or in that same respect in which it is spoken of as admitting of not being, either at a place or with respect to a size or to being of a certain sort; but what admits of not being with respect to thinghood is destructible simply. Therefore nothing that is simply indestructible is simply in potency (though nothing prevents it from being potentially in some particular respect, such as of a certain sort or at a certain place), and so all of them are at work. And none of the things that are by necessity is in potency (and yet these are primary, since if these were not, nothing would be), nor is motion, if there is any everlasting one; and if there is anything everlastingly in motion, it is not in potency to be moved other than from somewhere to somewhere (and nothing prevents material for this from belonging to it), and this is why the sun and the stars and the whole heaven are always at work and there is no fear lest they ever stop, which those who write about nature do fear. Nor do they grow weary in doing this, since for them motion does not concern the potency for a pair of contradictory things, as it does for destructible things,[18] so that the continuation of the motion would be

1050b 10

1050b 20

---

[17] Someone who has one of the potencies that involve reason is capable of a pair of *contrary* effects, which lie at opposite ends of a spectrum of possibilities, as a doctor can cause sickness as well as health; but every potency is capable of the contradictory effects of coming to be at work or failing to do so. Fire cannot cool anything, but external causes can prevent it from heating something.

[18] The everlasting motions of the stars are understood as circular, while motions on earth are not capable of continuing indefinitely without reversing, which means they must sometimes be going against the grain of the moving thing or the medium in which it moves. See *Physics* Bk. VIII, Ch. 8.

wearisome; for the cause of this is that the thinghood of destructible things is comprised of material and potency.

And things that undergo change, such as earth and fire, mimic the indestructible things, since they too are always at work, for they have motion in virtue of themselves and in themselves. But the other potencies, from which these have been distinguished, are all capable of contradictory things (for what is capable of moving something in this way is also capable of moving it not in this way); those at least that result from reason are of that kind, but with the irrational ones, the same potencies will be capable of contradictory effects in the sense of their being present or not. If, therefore, there are some such natures or independent things of the sort that people speak of in arguments about the forms, there would be something much more knowing than knowledge itself, and much more in motion than motion itself, since those things would have more the character of being-at-work, while the things they speak of are potencies for those.[19] So it is clear that being-at-work takes precedence over potency, and also over any source of change.

1050b 30

1051a

Chapter 9   And that being-at-work is a better and more honorable thing than a potency for something worth choosing, is clear from these considerations. For whatever is spoken of as being potential is itself capable of opposite effects; for instance, the same thing that is said to be potentially healthy is also, and at the same time, potentially sick, since the same potency belongs to being healthy and to running down, or to being at rest and to being in motion, or to building up and to knocking down, or to being built and to falling down. So being, potentially, opposite things belongs to something at one time, but the opposite things are incapable of belonging to it at the same time, and the ways of being at work are incapable of being present at the same time (such

1051a 10

---

[19] As always, one must distinguish a working out of an understanding of forms from a rejection of them. It is said in Plato's *Sophist* (248E–249B) that motion, life, and soul must belong to the forms, but this is never pursued in the dialogues. In Bk. I, Ch. 9, above (991a 23–24), Aristotle asked what it is that is at work, causing things to participate in the forms. He has just said, at 1050b 1–2, that the form *is* a being-at-work. And the usual way that commentators speak of a "Platonic" form that is separate from the particular things, and an "Aristotelian" form that exists only in the particulars, makes no sense either, since Aristotle has said that the form always pre-exists anything that comes into being (e. g., at 1034b 8), and since from 1043b 14 on, this inquiry has been on the level of form alone.

as being healthy and being run down), and so one of these must be the good one,[20] while the being-potential is equally both or neither; therefore the being-at-work is better. And in the case of bad things, it is necessary that the completion and being-at-work be worse than the potency, since the potential thing is itself capable of both opposites. Therefore it is clear that there is nothing bad apart from particular things, since the bad is by nature secondary to potency. Therefore

1051a 20   among things that are from the beginning and are everlasting, there is nothing bad, erring, or corruptible (since corruption is also one of the bad things).[21]

And geometrical constructions are discovered by means of activity, since it is by dividing up the figures that people discover them. If the figures were already divided, the constructions would be evident, but as it is, the constructions are present in the figures in potency. Why, say, does a triangle have two right angles? Because the angles around one point are equal to two right angles. So if the base were drawn beyond the side, this would be clear immediately to the one who looks at it.[22] And why is the angle in a semicircle always right? Because three lines are equal, the two halves of the base and the one erected from the center at right angles, which is clear to one who looks at it, who knows the other proposition just mentioned. And so it is clear that the things

1051a 30   that are in the figures in potency are discovered by being drawn into activity, and the cause is that contemplative thinking is the being-at-work of them; therefore the potency comes from a being-at-work, and for this reason it is only those who make a construction who know it

---

[20] The first sentence of the chapter eliminates indifferent things from consideration. So in the realm of things worth choosing, opposite things must have opposite weight with respect to choice.

[21] The words good, bad, better, and worse in this paragraph do not have a moral sense, as the examples show. There is nothing immoral about being sick, or about a house's falling down. These are things that are bad for the wholeness and characteristic being-at-work of some sort of being. Nature or art gives the standard of goodness. Moral goodness is derivative from goodness in this broader sense, since it is one of the things that fosters the emergence and well-being of human nature.

[22] The construction may be found in Euclid's *Elements* Book I, Proposition 32, but a second line has to be drawn to complete it. The proposition mentioned next is in Euclid (Book III, Proposition 31) but is not approached there in Aristotle's way, by means of the isosceles right triangle. In both cases, Aristotle gives too little information for a proof, but enough for someone fiddling with the figure to find a proof.

(since the being-at-work that belongs to a number is later in coming into being).[23]

Chapter 10   And since being and not being are meant in one sense by reference to the various ways of attributing being, and in another by reference to the potency or being-at-work of these or their opposites,   1051b
but the most governing sense is the true or the false, and since this as applied to particular things depends on combining and separating, in such a way that one who thinks that what is separated is separated and what is combined is combined thinks truly, but one who thinks these things to be opposite to the way the things are thinks falsely, when is what is spoken of as truth or falsity present or not present? For it is necessary to examine in what way we mean this. For you are not pale because we think truly that you are pale, but rather it is because you are pale that we who say so speak the truth.[24] So if some things are always combined and incapable of being separated, while others are   1051b 10
always separate and incapable of being combined, then others admit of opposites (for being something is being-combined and being-one, while not being it is not being combined but being more than one). Concerning the ones that admit of opposites, then, the same opinion and the same statement come to be false and true, and it is possible for someone at one time to think truly and at another time to think falsely; but about the ones that are incapable of being otherwise, nothing comes to be at one time true, at another time false, but the same things are always true and false.

---

[23] The last sentence is difficult to translate. The thought seems to be this: Mathematical things, according to Aristotle (1061a 29 and following), depend upon an act of abstraction, and this is true of them alone. Therefore they have no potency striving to emerge if nothing prevents it, and no being-at-work except in and dependent upon the activity of an intellect. Later thinkers such as Thomas Hobbes (e. g., in *On Body,* Ch. 1, Sec. 5) say that we know only what we make, and give mathematical examples. Aristotle agrees in the case of mathematics, but otherwise believes that the object of knowledge is independent and at work upon the knower.

[24] This is a dialectical step beyond what is said in Bk. VI, Ch. 4. Being as the true was there set aside as not the governing sense of being, since it belonged only to thinking and didn't reveal any new kind of being. Books VII–IX have revealed the being of everything in our sensory experience, and the dependence of it all upon the being-at-work of forms to make things whole. But that leads to the question of how being should be understood when there is no division into "categories," no potency, and no wholeness out of parts. This is a shift to a more primitive and original sense of truth as emergence out of hiddenness, which has been anticipated at the end of Bk. II, Ch. 1.

But now for things that are not compound, what is being or not being, and the true and the false? For the thing is not a compound, so

1051b 20   that it would be when it is combined and not be if it is separated, like the white on a block of wood or the incommensurability of a diagonal; and the true and the false will not still be present in a way similar to those things. Rather, just as the true is not the same thing for these things, so too being is not the same for them, but the true or false is this: touching and affirming something uncompounded is the true (for affirming is not the same thing as asserting a predication), while not touching is being ignorant (for it is not possible to be deceived about what it is, except incidentally). And it is similar with what concerns independent things that are not compound, since it is not possible to be deceived about them; and they are all at work, not in potency, for otherwise they would be coming into being and passing away, but the very thing that *is* does not come to be or pass away, since it would have to come

1051b 30   from something. So it is not possible to be deceived about anything the very being of which is being-at-work, but one either grasps or does not grasp it in contemplative thinking; about them, inquiring after what they are is asking whether they are of certain kinds or not.

So being in the sense of the true, and not-being in the sense of the false, in one way is: if something is combined, it is true, and if it is not combined, it is false. But in one way it is: if something *is,* it is present

1052a   in a certain way, and if it is not present in that way, it is not. The true is the contemplative knowing of these things, and there is no falsity, nor deception, but only ignorance, and not the same sort of thing as blindness; for blindness would be as if someone were not to have the contemplative power at all. And it is clear also that about motionless things there can be no deception about when they are so, if one grasps that they are motionless. For if one supposes that the triangle does not change, one will not think that it sometimes has two right angles and sometimes does not (for it would be changing); but it is possible that one such thing be a certain way and another not (for instance, that no even number is prime), or that some are and some are not. But about

1052a 10   a single number, not even this is possible, since one could no longer suppose that one thing is one way and another not, but one thinks truly or falsely that something is always a certain way.

Book X (Book I)
Wholeness[1]

Chapter 1 That oneness is meant in more than one way has been 1052a 15 said before, in the distinctions made about the various senses in which things are meant; and though it has more meanings, when the ways it is meant are gathered under headings, there are four senses in which something is said to be one primarily and in its own right, rather than incidentally. For oneness belongs to what is continuous, either simply or, especially, by nature, and not by contact or by a binding cord (and 1052a 20 among these that is more so one and is more primary of which the motion is more indivisible and more simple); and it belongs still more to what is whole and has some form and look, especially if something is of that sort by nature and not by force, as those things are that are so by means of glue or bolts or being tied with a cord, but rather has in itself that which is responsible for its being continuous. And something is of this sort if its motion is one and indivisible in place and time; and so it is clear that, if something has a source of motion that moves it in the primary kind of the primary class of motions (by which I mean the circular type of change of place), this is one magnitude in the primary sense. So some things are one in this way, insofar as they are continuous or whole, but others are one because the articulation of them is one, 1052a 30 and of this sort are those things of which the thinking is one, and this in turn is of this sort if it is indivisible, and an act of thinking is indivisible if it is of something indivisible in form or in number. Accordingly, a particular thing is one by being indivisible in number, but that which is one by means of intelligibility and knowledge is indivisible in form, so that what is responsible for the oneness of independent things would be one in the primary sense. Oneness, then, is meant in this many ways, the naturally continuous, the whole, the particular, and the universal,[2]

---

[1] The topic of the book moves from oneness to manyness to contrariety in general. The bulk of it thus supplies what is called for at 1003b 35–1004a 2, as germane to first philosophy but not central to it. Dialectically, though, its primary role is as a link between Books IX and XII, since its first chapter establishes the unity of anything that has a form as that of which the thinking is one, and its second chapter points on to the being that is the source of such wholeness. This title for Book X supplied by the translator.

[2] This last kind of oneness has to refer to what has just been said to be indivisible in

1052b

and each of these is one by being indivisible, either with respect to motion, or to an act of thinking, or in articulation.

But it is necessary to notice that one must not take the sorts of things that are spoken of as one as being meant in the same way as what it is to be one, or what the articulation of it is. For oneness is meant in all the ways mentioned, and each thing to which any of these ways belongs will be one; but being-one will sometimes belong to one of these, and sometimes to something else, which is even closer to the name, though these four senses are closer to the meaning of it. The same thing would also be the case with "element" and "cause," if one had to speak about them, distinguishing the things to which the words are applied, and

1052b 10

giving a definition of the words. For there is a sense in which fire is an element (though presumably in its own right the element is the indeterminate or something else of that sort), and a sense in which it is not; for being fire is not the same thing as being an element, but as a certain thing and a nature, fire is an element, and the word indicates that this attribute goes along with it because something is made out of it as its first constituent. And it is that way also with "cause" and "one" and all such things, and this is why being one is being indivisible, just exactly what it is to be a *this*, separate on its own in either place or form or thinking, or to be both whole and indivisible, but especially to be the primary measure of each class of things, and, in the most governing sense, of the class of things with quantity, for it has come from there to apply to other things.

1052b 20

For a measure is that by which the amount of something is known; and it is either by a one or by a number that an amount is known, insofar as it is an amount, while every number is known by a one, so that every amount, insofar as it is an amount, is known by that which is one, and that by which amounts are known first of all is the one itself; hence oneness is the source of number as number. And based on that, measure is spoken of in other areas too as that by which each

---

form and the object of a single act of thinking, and to have as its primary instance that which makes an independent thing one. It cannot, therefore, be a universal in the sense that corresponds to a general class, a common property, or anything that is one but applied to many things. Book VII, Ch. 13, eliminated the universal in that sense from consideration among the causes and sources of being as such. Such a cause must be a *this* and separate, independent of our thinking, and universal only in the sense that it is responsible for all the particulars of some definite kind. For this sense of "universal," see 1026a 29–32.

thing is primarily known, and of each sort of thing, the measure is a
one, in length, in breadth, in depth, in weight, in speed (for weight
and speed are applied in common to contraries, since each of them
is meant in two ways, as weight is applied both to that which has
any downward inclination whatever and to that which has an excess
of downward inclination, and speed is applied both to that which has   1052b 30
any motion whatever, since it has some speed, and to that which has an
excess of motion, and even the lightest weight has some weight). In all
these cases, the measure and source is something one and indivisible,
since even among lines, one uses the foot as though it were indivisible.
For everywhere, people seek as a measure something that is one and
indivisible, and this is what is simple either in kind or in amount. So
wherever it seems not to be possible to take away or add anything,
there is an exact measure, and that is why the measure of number is   1053a
the most exact, since people set down the numerical unit as indivisible
by anything. And in other cases, the measures are imitations of this
sort of measure, for from a mile or a ton, or always from some very
large thing, it escapes notice if something has been added or taken
away, more than from something smaller; and so the first thing which,
according to perception, does not admit of being added to or subtracted
from, all people make into a measure, of wet things or dry things or
weight or extent, and then think they know the amount of something,
whenever they know it by this measure.

  And people measure even motion by the simple motion and by the
fastest (since this takes the least time), and this is why in astronomy   1053a 10
something that is one in this way is a source and measure (for people
assume that the motion of the heaven is uniform and fastest, and judge
the others in relation to it), and in music it is the quarter tone, because
it is the smallest interval, and in speech it is the letter. And each of these
is one in the same way, not in a single sense common to them all, but
just in the way mentioned. But the measure is not always numerically
one, but sometimes more than one, as there are two types of quarter
tone, distinct not for hearing but in their ratios, and the sounds by
which we measure speech are more than one, and the diagonal and
side of a square have two measures, and similarly with all magnitudes.
So in this way the measure of everything is what is one, because we
recognize what the thinghood of a thing consists of by dividing it either
in amount or in form. And for this reason the one is indivisible, because   1053a 20
the first thing by which each kind is known is indivisible, but they are

not all indivisible in the same way, for instance a foot in length and a unit in arithmetic, but the latter is not divisible in any way, while the former presents itself among things indivisible as far as perception is concerned, as was said before, since presumably every continuous thing is divisible.

And a measure is always the same kind of thing as what it measures, for the measure of magnitudes is a magnitude, and in particular, that of length is a length, of breadth a breadth, of spoken sounds a spoken sound, of weight a weight, and of numerical units a numerical unit. And it is necessary to put the last example this way, and not to say that a number is the measure of numbers, though that would have been appropriate if it were analogous to the other cases; but one would not be regarding it in an analogous way, but rather just as if one were to
1053a 30   regard units, and not *a* unit, as the measure of units, since a number is a multitude of units.

And we speak of knowledge or sense perception as a measure of things for the same reason, because we recognize something by means of them, although they are measured more than they measure. But what happens to us is just as if, after someone else had measured us, we recognized how big we are by the ruler's having been held up to us so many times. And Protagoras says a human being is the
1053b   measure of all things, as if he were saying that a knower or perceiver were the measure, and these because the one has knowledge and the other perception, which we say are the measures of their objects. So while saying nothing, these people appear to be saying something extraordinary.

So it is clear that being one, for someone defining it most strictly in its literal sense, is being a certain kind of measure, of an amount in the most governing sense of oneness, and next of a kind; and something will be of this sort if it is indivisible in amount, and something else if it is indivisible in kind. For this reason what is one is what is indivisible, either simply or in the respect in which it is one.

Chapter 2   As for the thinghood and nature of what is being
1053b 10   inquired about, and its manner of being, we made an approach in the discussion of impasses to the question what oneness is, and how one ought to understand it, whether as though the one itself is a particular independent thing, as the Pythagoreans said in earlier times and Plato later, or whether instead some nature underlies it, and it ought in some

way to be explained in more familiar terms and more in the manner of those who write about nature; for a certain one of them says that oneness is love, another that it is air, and still another that it is the unbounded.[3]

Now if none of the universals can be an independent thing, as was said in the chapters about thinghood and about being, since not even being itself is an independent thing as though it were some one thing capable of having being apart from the many beings (since it is common to them), other than solely as a thing attributed to them, it is clear that oneness is not a universal either; for being and oneness are predicated in the most universal way of all things. And so, general classes are not particular natures and independent things separate from the rest, nor could oneness be a general class, for the same reasons on account of which neither being nor thinghood can be a general class of things.[4]

1053b 20

What's more, what is true about oneness must hold true in a similar way for all things; and being and oneness are meant in equally many ways. Therefore, since among qualities there is a certain kind of oneness and some nature that is one, and similarly also among quantities, it is clear that also in general one must inquire after what oneness is, just as also what being is, since it is not sufficient just to say that it is itself its own nature. And surely among colors what is one is a color, such as white, and then the other colors clearly come to be out of this and black, black being the deprivation of white just as darkness is the deprivation of light; and so, if beings were colors, the things that are would be a certain kind of number, but of what? It is obvious that they would be a number of colors, and oneness would be a certain *kind* of unity, such as whiteness. And similarly, if beings were melodies, they would be a certain number, though in this case a number of quarter tones, but the thinghood of them would not be a number; and oneness would be something of which the thinghood was not oneness but the quarter tone. And similarly again in the case of spoken utterances, beings would be a number of letters, and oneness

1053b 30

1054a

---

[3] In the first case (Empedocles), love is the active source that makes things combine into one, while in the next two (Anaximenes and Anaximander) what is named is the one material source out of which all things arise.

[4] The argument for this, at 998b 22–28, is put in a technical way, but amounts to this: an all-comprehensive class could not have subclasses, since they would have to be distinguished by having or not having some characteristic *outside* the original class.

of which it is also in between. And what is neither good nor bad is opposed to both, but is nameless, for each of those two is meant in a number of ways, and there is not just one thing receptive of them; this last is more the case with what is neither white nor black, but not even this is meant in one sense, though the things to which this negation is applied as a deprivation are somehow limited, since they have to

1056a 30   be either gray or yellow or something else of that sort. So those who consider all things to be meant in the same way do not estimate them rightly, since what is neither a shoe nor a hand would be between a shoe and a hand, because what is neither good nor bad is between good and bad, as though there were some sort of in-between in everything. But it is not necessary that this follow. For the joint negation of opposites is present for things which are of such a nature that there is some sort

1056b   of in-between and interval. But between things of the other sort there is not a difference, since the joint negations are of things of which each is in a genus other than of the other, so that the underlying subject is not one thing.

Chapter 6 In a similar way, one might be at an impasse about the one and the many. For if the many are simply opposite to the one, some impossible things follow. For the one would be few, or a few, since the many are also opposite to the few. Also, two would be many, if what is twofold is spoken of as manifold but corresponds to two, so that the one is few; for in relation to what are two many if not to the one

1056b 10   and the few? For nothing is fewer. Also, if as the long and the short are in length, so are the many and the few in multitude, and what is many is much (unless they differ somehow in something continuous with fluid boundaries), the few would be a certain kind of manyness. Therefore the one would be a certain kind of manyness, if it is also few, and this is necessary if two are many.

But perhaps the many are said in a certain way to be also much, but in a different sense; for example, water is much but not many. But many is said of all those things that are divided, in one way if there is a multitude having an excess either simply or in relation to something (and the few in the same way means a multitude having a deficiency), in another way as a number, in which sense alone it is also opposite to

1056b 20   the one. For in this way we speak of one and many in just the way one might say one and ones, or a white thing and white things, or speak of the things measured off in relation to their measure; in this way too, manifold things are spoken of, for each number is many because it consists of ones and because each number is measured by the one, and is many as opposed to the one and not to the few. In this sense, then, even two are many, but this is not as a multitude having an excess either in relation to anything or simply, but as the first multitude. But in the simple sense, two are few, for this is the first multitude of those that have a deficiency. (And this is why Anaxagoras was not right to leave off after saying that all things together were infinite both in multitude and in smallness, though he ought to have said, instead of "and in smallness," "and in fewness," but that is not infinite.) For things are few not on account of the one, as some say, but on account of the two.

1056b 30

The one is opposed to the many in numbers as a measure to the thing measured, and these are opposed as relative terms, but as things which do not in their own right belong among relative things. And it has been distinguished by us elsewhere[12] that relative is meant in two ways, some things being relative as contraries, others in the manner of knowledge in relation to the thing known, because of something else's being spoken of in relation to it. But nothing prevents the one from being less than something, such as two, since if it is less it is not also few. But multitude is as though it were the general class of number, for a number is a multitude measured by the one, and the one and number are opposed in a certain way, not as contrary but in the way some relative terms are said to be opposed; for insofar as something is a measure and something else is measured, in this respect they are opposed, for which reason not everything that is one is a number, for instance if something is indivisible. But even though knowledge is spoken of in a similar way in relation to the thing known, it does not yield a similar result, for while knowledge might seem to be the measure, and the thing known what is measured, it turns out that, while everything that is known is knowable, not every knowable thing is known, because, in a certain sense, knowledge is measured by the thing known.

1057a

1057a 10

Multitude is contrary neither to the few—to this the many are contrary as a multitude that exceeds is contrary to what is exceeded in multitude—nor to the one in every sense. But in one sense they are contrary, as was said, because multitude is divided and the one

---

12 See Bk. V, Ch. 15.

is indivisible, though in another sense they are opposed as relative terms, as are knowledge and the thing known, if multitude is number and the one is the measure.

Chapter 7 Since contraries admit of having something in-between, and some do have it, it is necessary for what is in-between to be made out of the contraries. For all in-between things are in the same general class as the things they are between. For we speak of as in-between those things into which something that changes must change first (for instance, if one were to change from the highest to the lowest tone of a chord by steps of the least interval, one would come first to the in-between sound, or in colors if one were to change from white to black, one would come to red or gray before black, and similarly in other cases), but to change from one general class into another is not possible other than incidentally, such as from a color into a shape. Therefore it is necessary for the in-between things themselves to be in the same general class and in the same one as the things they are between.

But surely all in-between things are in-between some sort of opposites, for only from these is it possible for something to change in its own right (for which reason it is impossible to be between things that are not opposites, since then there could be a change that was not from opposites). But among opposites, between contradictories there can be no in-between (for this is contradiction: an opposition in which one of the two parts must be present to anything whatever, which thus has nothing in-between), and of the rest, some are relative terms, some deprivations, and others contraries. Among relative terms, those that are not contraries do not have an in-between, and the reason is that they are not in the same general class. For what is in-between knowledge and the thing known? But between the great and the small there is something. But if in-between things are in the same general class as the things they are between, as has been shown, and are between contraries, it is necessary that they be composed out of these contraries. For there will be either some general class that includes them, or none. And if there is a general class of such a kind that it takes precedence over the contraries, the specific differences that make those contraries species of a genus will be contraries of a prior sort, since the species are composed of the genus and the specific differences. (For example, if white and black are the contraries, and one is the dilating color,

1057a 20

1057a 30

1057b

the other the contracting color,[13] these specific differences, dilating   1057b 10
and contracting, are prior to white and black, so that these are contrary to one another in a prior way.) For surely the things that differ
contrarily are more contrary, while the remaining contraries and the
things in-between them are derivative from the genus and the specific
differences. (For example, as many colors as are between white and
black must be said to be derived from the genus—and the genus is
color—and certain specific differences, but these latter will not be the
primary contraries, for if they were, every color would be either white
or black. So they are different, and therefore they will be between the
primary contraries; but these primary specific differences are dilating
and contracting.)

So one must inquire first, about those contraries that are not in   1057b 20
a general class, what it is of which the things in-between them are
composed (for things in the same genus must be composed of things
that are not compounded with the genus or else be uncompounded).
Now contraries are not composed of one another, and are therefore
sources, while the things between them are either all composed of
them, or none. But something comes to be out of the contraries, in such
a way that a change will be into this before it is into the other contrary,
for it will be more of one of them and less of the other. Therefore this
also is in-between the contraries. And therefore all the other in-between
things are composite, for a thing that is composed of the more and the
less is in some way derived from those things with respect to which it
is said to be more and less. And since there are no other things prior
to them that are the same in kind as the contraries, all the in-between   1057b 30
things would be derived from the contraries, so that also all the lower
kinds, both contraries and in-between things, would be derived from
the first contraries. That, then, all in-between things are in the same
general class as, and are between, contraries, and are all composed of
those contraries, is clear.

Chapter 8   That which is other in species is a certain kind of thing
that is other than something, and it is necessary that this kind belong
to both things; for instance, if an animal is other in species, they are
both animals. Therefore it is necessary that things other in species be

---

13 This is the likely story Timaeus tells about color in Plato's *Timaeus,* 67 D–E.

1058a

in a genus that is the same; for that one thing which is said to be the same in both, not having a specific difference that is incidental, whether it is present as material or in another way, I call a genus in this sense.**14** For not only is it necessary that there be a common thing present, as, say, both are animals, but also this same animalness must be other in each of the two, as, say, with the one a horse, with the other a human being, on account of which this common thing is other in species for each than for the other. So the one will be in its own right a certain sort of animal, the other in its own right a certain sort, such as the one a horse, the other a human being. Therefore it is necessary that this difference be an otherness that belongs to the genus; for by a difference that belongs to the genus I mean that which makes this same thing be other. This, then, will be a contrariety (and this is clear

1058a 10

also from examples), for all things are divided by opposites, and it has been shown that contraries are in the same genus. For contrariety is complete difference, and every difference in species is a certain kind of thing that is other than something, so that this same thing is also the genus in both things. (This is why all the contraries that are different in species but not in genus are in the same list under the ways of attributing being,**15** and are the most different from one another, since the difference is complete, and do not come into being together with one another.) Therefore the specific difference is a contrariety.

This, therefore, is being other in species: the contrariety that things in the same genus, that are indivisible, have (while those that do not have a contrariety, which are indivisible, are the same in species).**16**

1058a 20

For in divisions contrarieties come about also in the in-between things before one comes to the indivisible ones; so it is clear that, in relation to what is called the genus, none of the species is either the same as it or different from it in species (and fittingly so, for material is made evident by negation, and the genus is the material of that of which it is said to be the genus—not the genus in the sense of the generations of descendants of Heracles, but in the sense that is in the nature of a

---

14 See 1054b 32–33 and note.

15 See the footnote to 1054a 31.

16 Within a species, what is indivisible is the individual or particular thing; within a genus, what is indivisible is the species understood as a kind that is unmixed with its opposite.

thing), nor is any of them the same or different in species in relation to the things not in the same genus, but they will differ from them in genus, but in species from things in the same genus. For it is necessary that the difference of things that differ in species be a contrariety, and this belongs only to things which are in the same genus.

Chapter 9 One might be at an impasse as to why woman does not differ from man in species, since female and male are contraries and their difference is a contrariety, nor are a female and a male animal different in species, even though this difference belongs to an animal in its own right and not in the way that white or black do, but it is insofar as it is an animal that female and male belong to it. And this impasse is just about the same as why one pair of contraries makes things different in species but another does not, as footed and winged do while white and black do not. Is it that the one pair are attributes fitting the nature of the genus while the other are less so? And since one sort of thing is articulation while another is material, those contrarieties that are in the articulation make a difference in species, but those that are in what is conceived together with the material do not make such a difference. This is the reason whiteness of a human being, and blackness, do not make such a difference, nor is there a difference in species of a white human being in relation to a black one, even if one were to set down one name for each. For a human being is as material, and material is not what makes a difference; for human beings are not on this account species of human beings, even though the flesh and bones of which this one and that one are made are different, but it is the composite whole that is different, though not different in species, because there is no contrariety in the articulation.

This is the ultimate indivisible thing, and Callias is the articulation with material, and so it is a white human being, because Callias is white; thus the human being is white incidentally. Neither are a bronze and a wooden circle different in species; even a bronze triangle and a wooden circle do not differ in species on account of their material, but because there is a contrariety present within their articulations. But does the material not make things different in species when it is different in a certain way, or is there a sense in which it does make them different? For why is this particular horse different in species from this particular human being, even though their articulations are present with material? Or is it because the contrariety is present in

1058a 30

1058b

1058b 10

1058b 20

the articulation? For also there is a difference between a white human being and a black horse, and it is a difference in species, but that is not insofar as the one is white and the other is black, since even if both were white they would still be different in species. And male and female are attributes fitting the nature of an animal, but not in its thinghood, but in its material and its body, on account of which the same germinal material becomes female or male by undergoing a certain way of being acted upon. So what it is to be different in species, and why some things differ in species and others do not, have been said.

1058b 30

Chapter 10   But even though contraries are different in species, while the destructible and indestructible are contraries (since a deprivation is a determinate incapacity), a destructible thing and an indestructible thing must be different in genus. Now, though, we have been speaking of the things themselves of which the names are universals, so it might not seem necessary that every indestructible and destructible thing whatever be different in species, just as neither are every white and black thing (since the same thing admits of being both, even at the same time, if it is one of the universals, as human being may be both white and black, and even if it is one of the particulars, since the same human being, though not at the same time, may be both pale and swarthy, even though white is contrary to black). But while some contraries belong to some things incidentally, such as those now being spoken of and many others, other contraries cannot, among which are the destructible and the indestructible; for nothing is destructible incidentally, since what is incidental admits of not being present, but destructibility is one of the things that are present by necessity in the things to which they belong. Otherwise one and the same thing would be destructible and indestructible, if destructibility admits of not belonging to it. Therefore it is necessary that destructibility either be the thinghood or be present in the thinghood of each destructible thing; and the same argument also concerns the indestructible, since both are among things present by necessity. Therefore, that in respect to which and as a result of which one thing is destructible and another is indestructible in the first place has an opposition, so that they must be different in genus.

1059a

1059a 10

It is clear then that it is not possible for there to be forms of such a sort as some people say they are,[17] for there would be a human being, one indestructible but another destructible. And yet the forms are said to be the same in species with the particulars, and not by an ambiguous name; but things which are different in genus stand further apart than those that differ in species.

---

[17] See 990a 34–990b 17, and footnotes.

Book XI (Book K)
Order[1]

Chapter 1 That wisdom is some sort of knowledge concerning the sources of things is clear from the first passages in which impasses were gone through about the things said by others about the sources,[2] but one might be at an impasse whether one ought to conceive of wisdom as one or more than one kind of knowledge; for if it is one, one kind of knowledge is always of contraries, but the sources of things are not contraries, but if it is not one, what sorts of knowledge ought one to set these down as being? Also, does it belong to one or to more than one kind of knowledge to contemplate the starting points of demonstration? If to one, why that one rather than any other? If to more than one, what sorts ought one to set these down as? Also, does wisdom concern all sorts of thinghood or not? For if it is not about all, it is hard to give an account of which ones it concerns, but if one knowledge concerns them all, it is unclear how it is possible for the same knowledge to be about more than one topic. Also, is it about the kinds of thinghood only, or also about their attributes? For if demonstration is possible concerning the attributes, about the kinds of thinghood it is not possible, but if the two kinds of knowledge are different, what is each and which is wisdom? For insofar as it is demonstrative, wisdom would be about attributes, but insofar as it is about primary things, it would be about the kinds of thinghood.

1059a 20

1059a 30

But neither ought one to set down the kind of knowledge being sought as concerning the causes spoken of in the writings about nature, since it is not about that for the sake of which (for this sort of cause is the good, and this belongs among actions and things that are in

---

[1] This title for Book XI supplied by the translator.

[2] This refers to Bk. III, Ch. 2. The first eight chapters of Bk. XI summarize the preliminary argument of the *Metaphysics,* from Bks. III, IV, and VI. The last four chapters of Bk. XI summarize parts of the *Physics,* from Bks. II, III, and V. Together they examine and refute an array of reasons why one might believe there is some sort of ultimate disorder in things. Is there such a thing as wisdom or unchanging being at all (Ch. 1–2)? Is knowledge of a number of unconnected kinds (Ch. 3–4, 7)? Is truth contradictory or relative (Ch. 5–6)? Are incidental being and chance events at the root of things (Ch. 8)? Is the world infinite (Ch. 10)? Are motion and change random or indeterminate (Ch. 9, 11–12)?

motion, and it moves things first—for that is the sort of thing an end is—but a thing that first moves them is not present among immovable things). And in general, there is an impasse whether the knowledge now being sought is about perceptible independent things at all, or not, but about other things. For if it is about others, it would be about either the forms or the mathematical things, but it is apparent that there are no forms. (Nevertheless, there is an impasse, even if one posits that there are forms, why in the world it is not the same with the other things of which there are forms as it is with the mathematical things; I mean that they place the mathematical things between the forms and the perceptible things as a third sort besides the forms and the things here, but there is no third human being or horse aside from the human or horse itself and the particulars, and if in turn it is not as they say, about what sort of things ought one to set down the mathematician as being busy with? For it is surely not about the things here, since none of these is the sort of thing that the mathematical kinds of knowledge inquire into.) But neither is the knowledge now being sought concerned with mathematical things (since none of them is separate), nor is it a knowledge of perceptible independent things, since they are destructible.

1059b

1059b 10

And in general, one might be at an impasse as to what sort of knowledge it is that considers impasses about the material of mathematical things. For neither is it the study of nature, since the whole business of one who studies nature is concerned with things that have in themselves a source of motion and rest, nor is it the examination of demonstration and knowledge, since it makes its inquiry about just that class of things. It remains, then, for the sort of philosophy that lies before us to make the examination about them. But one might be at an impasse whether one ought to set it down that the knowledge being sought concerns the sources that are called elements by some people, for everyone sets these down as ingredients in compound things. But it might seem instead that the knowledge being sought has to be about things that are universal, since every articulation and every sort of knowledge is about universals and not about ultimate particulars, and so that which concerns the primary classes of things would be that way too. These would turn out to be being and oneness, for these most of all might be supposed to include all beings and to be most like sources on account of being primary in nature, since if these were destroyed everything else would also be taken away along with them,

1059b 20

1059b 30

since everything *is* and is one. But insofar as it is necessary for their own specific differences to have a share in them, if one sets them down as general classes, while no specific difference has a share in its genus,[3] on this account it would seem that one ought not to set them down as general classes or as sources. Also, if the simpler thing is more a source than is the less simple thing, while the lowest classes cut off from the genus are simpler than those general classes (for they are indivisible, but the general classes are divided into a number of differing species), the species would seem to be sources more than would the general classes. But insofar as the species are destroyed along with the general classes, the general classes seem more like sources, for the thing that causes things to be destroyed along with it is a source. So the things that have impasses are these and others like them.

1060a

Chapter 2   Also, ought one to set down anything besides the particular things or not, and is the knowledge being sought about particulars? But these are infinite. Yet surely the things apart from the particulars are general classes or species, but the knowledge now being sought is not about either of these. Why this is impossible has been said.[4] For in general there is an impasse as to whether one ought to assume there to be any separate independent thing besides the perceptible independent things here around us, or not, or that these are the beings and wisdom is concerned with them. For we seem to be seeking something else, and this is what lies before us: I mean, to see whether there is anything separate by itself and belonging to none of the perceptible things. Also, if besides the perceptible independent things there is any other independent thing, alongside which of the perceptible things ought one to set this down as being? For why would one set it alongside human beings and horses more than alongside the

1060a 10

---

[3] There is no general class of all things that *are* or are one, since nothing would be left outside it to be specific differences and divide it into subclasses. There are, according to Aristotle in this work (1017a 23–27), eight highest irreducible classes of beings. See also 998b 22–28.

[4] Book VII, Ch. 16, is perhaps the place where this is made clearest. The species (the form understood as a universal), and even more so any more general universal classes, have no separate being and could not be causes. But since the particular things are unknowable, the central impasse in the way of the knowledge being sought is established. This knowledge must be of forms understood as separate (1017b 27–28) and as at-work (991a 23–24).

other animals and even the things without souls in totality? But surely to set up other everlasting things equal in number to the perceptible and destructible independent things would seem to fall outside the bounds of what is reasonable. But if the source now being sought

1060a 20   is not separate from bodies, what else could one posit other than material? Yet this does not have being as something at-work, but as in potency. And a more ruling source than this would seem to be the look or the form, but this is destructible,[5] so that there is no everlasting independent thing separate in its own right at all. But this is absurd, for it is evident that there is some such source and independent thing and it is the sort of thing sought for by the most refined thinkers as something that has being; for how would there be order if there were not something everlasting and separate and constant?

Also, if there is an independent thing and source of such a nature as we are now seeking, and this is one and the same for all things, both the everlasting and the destructible ones, there is an impasse as to why

1060a 30   in the world, when the source is the same, some of the things ruled by the source are everlasting but others are not everlasting (for this is absurd); but if there is one source of the destructible things and another of the everlasting ones, and if the source of the destructible things is everlasting, we are similarly at an impasse. (For why, if the source is everlasting, are the things ruled by the source not also everlasting?) But if it is perishable, there would turn out to be some other source of this one, and yet another of that one, and this would go on to infinity.

But if in turn one were to set down what seem most of all to

1060b   be unchanging sources, being and oneness, first, if each of them does not signify a *this* and an independent thing, how would they be separate and by themselves? But we are seeking everlasting and primary sources of that kind. If however each of them does indicate a *this* and an independent thing, all beings are independent things, since being is attributed to them all (and also oneness to some of them). But it is false that every being is an independent thing.[6] And also, as for those who say that the first source is the one, and generate number first from

1060b 10   the one and material, how is it possible for what they say to be true?

---

[5] If the form is understood merely as the ordering *of* a particular thing and *in* its material, it is destructible. See 1039b 20–25.

[6] Recall that beings include white, walking, yesterday, four ounces, and everything that is.

For how ought one to think of two, and each of the rest of the numbers that are composite, as one? For neither do they say anything about this, nor is it easy to say anything. And if one sets down lines or the things following from these (I mean the first surfaces) as sources, these are surely not separate independent things, but cuts and divisions, the former from surfaces and the latter from solids (but points from lines), and also they are limits of these same things, and all these are present in other things and none is separate. Also, how is one to understand there to be thinghood of a unit or a point? For of every independent thing there is a process of coming into being, but of a point there is not, since the point is a division.[7]

And there is besides an impasse, that all knowledge is of universals   1060b 20
and of the suchness of things, but thinghood does not belong to universals, but rather an independent thing is a *this* and is separate, so if there is knowledge about the sources, how ought one to understand the source to be an independent thing? Also, is there anything besides the composite whole (I mean the material and what is together with this), or not? For if not, then the destructible things in material are all the things there are, but if there is something else, this would be the look and the form; now this is in some cases difficult to mark off and in others not, for in some cases, such as that of a house, it is clear that the form is not separate. Also, are the sources the same in species or in number? For if they are one in number, all things would be the same.   1060b 30

Chapter 3 Now since the knowledge that belongs to the philosopher concerns being as being universally and not in relation to a part, while being is meant in a number of ways and not in a single sense, then if it is meant ambiguously and in accordance with nothing that is common to those meanings, it is not possible that it be subject to one knowledge (since there is not one class of such things), but if it is meant in accordance with something common, it would be subject to one knowledge. It seems to be meant in the way that has been spoken of, in just the way that medical and healthy are meant. For each of these is meant in a number of senses, but each of them is meant in this way: 1061a
by leading back in some way, in the one case to medical knowledge,

---

[7] This argument applies only to things that sometimes are and sometimes are not. See 1002a 30–1002b 11.

and in the other case to health, and though these lead back in various ways, in each case it is to the same thing. For a medical discourse or a medical knife are meant by the former's being from medical knowledge and by the latter's being useful for this. And it is similar also with healthy, for one thing is called so because it is a sign of health, another because it tends to produce it. And it is the same way also with the rest of its meanings. So it is also in the same way that being is meant in every instance, for it is by being an attribute of being as being, or an active state of it, or a disposition, or a motion of it, or something else of

1061a 10   that sort that each of them is called a being. And since for every being there turns out to be a leading back to some one thing that is common to them, then also each of the pairs of contraries will be traced back to the first differences and contrarieties of being, whether many and one or likeness and unlikeness are the first differences within being, or some other pair, for let these stand as having been examined. And it makes no difference whether the tracing back of being turns out to be toward being or toward oneness, for even if these are not the same but different, still they are interchangeable, for it is the case both that what is one also in some way *is,* and that what *is* is one.

　　And since it belongs to a knowledge that is one and the same to
1061a 20   study every pair of contraries, and each of them is spoken of by means of some deprivation (even though someone might raise an impasse about some of them as to how they are meant in accordance with a deprivation when there is something in-between them, as in the case of the unjust and the just, so that in all such cases one must set down that the deprivation is not a lack of the whole articulation, but of the extreme form of it, for example, if the just person is by some active condition obedient to the laws, the unjust person would not be lacking the whole of that articulation in every respect, but failing in some respect concerning obeying the laws, and in this sense the deprivation will belong to him, and in the same way also in the other cases), and just as the mathematician makes his study about things
1061a 30   that result from taking something away (for he studies things after having stripped away everything perceptible, such as heaviness and lightness, hardness and its opposite, and also hotness and coldness and the other pairs of contrary perceptible attributes, and this leaves behind only what is of some amount and continuous, belonging to some things in one dimension, some in two, and others in three, and he studies the attributes that belong to these insofar as they are of some

amount and continuous, and in no other respect, and he examines the
relative positions of some, and the things that follow from these, and
of others the kinds of commensurability and incommensurability, and
of others the ratios, but still we set it down that there is one and the
same knowledge of all this, namely geometry), so too is it the same
way with being (for the attributes that belong to this to the extent that
it is being, and the contrarieties within it insofar as it is being, belong to
no other sort of knowledge than philosophy to study, since one would
assign the study of them to physics not insofar as they *are*, but rather
insofar as they have a share in motion, while dialectic and sophistry
are concerned with the attributes of beings, but not insofar as they are,
and they are not concerned with being itself to the extent that it is) so
that it remains for the philosopher to be the one who studies the things
that have been spoken of, to the extent that they are beings.

1061b

1061b 10

Since all being is meant in accordance with something that is one
and common, even though it is meant in a number of ways, and it is
the same way with the contraries (since they are traced back to the first
contrarieties and differences of being), such things are capable of being
subject to one knowledge, and the impasse spoken of at the beginning
would be resolved—I mean that in which it was raised as an impasse
how there could be one knowledge that concerns beings that are many
and different in kind.

Chapter 4   And since the mathematician uses common notions
in a particular way, it would also belong to the primary sort of
philosophy to study the things that govern these. For that, when equals
are subtracted from equals, the remainders are equal, is common to
all things that have amounts, but mathematics makes its study by
cutting out around its boundary a certain part of the material that is
appropriate to it, such as around lines or angles or numbers or any of
the other quantities, not insofar as they *are* but insofar as each of them is
continuous in one, two, or three dimensions; but philosophy does not
make its examination about a part of things, insofar as some attribute
belongs to each of them, but studies being, insofar as each such part *is*.
And it is the same way also with the knowledge about nature as with
mathematics, for physics studies the attributes and sources of beings
insofar as they are in motion and not insofar as they *are* (but we have
said that the primary sort of knowledge is about these things to the
extent that the things underlying them are beings, but not insofar as

1061b 20

1061b 30

they are anything else). For this reason one must set down both this sort of knowledge and the mathematical sort as parts of wisdom.

Chapter 5   There is a certain principle in beings about which it is not possible to be mistaken, but always necessary to do the opposite, and I mean to attain the truth, and it is of this sort: that it is impossible

1062a   for the same thing at one and the same time to be and not to be, and for the other things that are opposite to each other in that way to belong to it. And about such things there is no demonstration simply, but in relation to some person there is; for it is not possible to make a syllogism to this from a starting point more secure than it, but this is necessary if there is to be a demonstrating of it simply. But in relation to a person who makes opposite statements, it is necessary for the one showing why he is wrong to get from him something of a sort that will be the same as "it is impossible for the same thing to be and not be at one and the same time," but will not seem to be the same, for

1062a 10   only in this way can it be demonstrated to the one who says that the opposed statements can be true of the same thing. Now those who are going to participate in a discussion with each other must in some way understand what they say, for if this does not happen, how will there be a participation of these people in a discussion with each other? It is necessary then for each of the words to be intelligible and to mean something, and not many things but only one, but if it does mean more than one thing, it is necessary to make clear to which of these one is applying the word. So the one who says "this is and is not" denies that which he says, and so he denies that this word means what it means, which is impossible. So if being-this means something, it is impossible for the contradiction to be true.

1062a 20   Also, if the word means something, and if this is true of something, it must be this by necessity, but that which is so by necessity does not admit of sometimes not being so; therefore it is not possible for the opposed statement and contradiction to be true of the same thing. Also, if the statement were no more true than its contradiction, the one saying "human being" will be no more correct than one saying "not a human being," but it would seem that in saying a human being is not a horse, one is either more, or at least not less, correct than one who says a human being is not a human being, but the result is that one would also be correct in saying that the same person is a horse (since the opposite things were to be alike true); it follows then that

the same human being is also a horse or any of the other animals. 1062a 30
So while there is no demonstration of these things simply, surely in
relation to the one who posits these contradictory things, there is a
demonstration. And one who questioned even Heracleitus himself[8] in
this way would have quickly compelled him to agree that opposite
statements are never capable of being true of the same things. But as it
is, he took hold of this opinion without understanding what his own
words meant. But in general, if the thing said by him is true, not even
that itself would be true—I mean, that it is possible for the same thing 1062b
at one and the same time both to be and not be; for just as, when they
are separated, the assertion would be no more true than the denial, in
the same way too, since the pair of them put together and intertwined
are just like some one assertion, the whole as put in an assertion will
be no more true than its denial. Also, if nothing is truly asserted, even
this itself would be false—to assert that there is no true assertion. But if
there is any, that would refute the thing said by those who make such 1062b 10
attacks which completely abolish conversation.

Chapter 6 Something closely resembling the things being discussed is what was said by Protagoras, for he said that a human being
is the measure of all things, meaning nothing else than that what seems
so to each person is solidly so; but if this becomes so, it follows that
the same thing both is and is not, and is both bad and good, and the
rest of the statements asserted as opposites follow, since it often seems
to some particular people that this thing is beautiful, but it seems the
opposite to others, while the measure is the thing that seems so to each. 1062b 20
This impasse can be resolved by those who examine the source from
which this judgement has been taken, for it seems to have come about
for some people from the opinion of those who study nature, and for
others from the fact that not all people discern the same things about
the same things, but this thing seems sweet to some people and the
opposite to others. For it is a common teaching of just about all those
who have concerned themselves with nature that nothing comes to be
out of what is not, but everything out of what is; so since what is not
white comes to be out of what is completely white and in no way not
white, the thing that had become white could come to be out of what

---

[8] See 1005b 24–26, and footnote.

1062b 30

1063a

1063a 10

was not white, so that, according to them, it would come to be out of what was not unless in the first place the same thing was white and not white. But it is not difficult to resolve this impasse, for it has been said in the writings about nature in what way a thing that comes to be comes from what is not and in what way from what is.[9] And surely, when people are in dispute with one another, to hold alike to the opinions and the way things seem on both sides is silly, for it is clear that those on one of the two sides must be in error. And this is evident from what happens in sense perception, for never does the same thing seem sweet to some and the opposite to others unless the sense organ and means of judging the flavors mentioned is disabled and damaged in those on one of the two sides. And if this is so, one ought to assume that those on one side are the measure and not to assume this of the others. And I mean this similarly in the case of good and bad, and of beautiful and ugly, and all the other such things; for it is no different to give credence to this or to what appears to those who push a finger under their eyes and make out of one thing an appearance of two, and so think that there have to be two things because they appear that many, and then again one, for to those who do not move their eyes, one thing appears to be one thing.[10]

And in general it is absurd to make a judgement about the truth based on the fact that the things around us obviously change and never remain in the same condition, for in order to hunt for the truth, one ought to base one's judgement on the things that always hold on in the same condition and never make a change, and the things that pertain to the cosmos are of that sort, for these do not appear at one time in a certain way but then in another way, but always the same and not taking part in any change. Also, if there is motion and something is

---

[9] Everything that comes to be comes from what *is* in some way but is not what it will become, or is in potency what it will become but not so at-work. The error of Parmenides, that change and manyness are impossible, and various partial versions of the same error, are traced by Aristotle, in Bk. I, Ch. 8, of the *Physics,* to the failure to make these distinctions. See also 1032b 31–32.

[10] See 1010b 2–30, where this is discussed more fully, and the references given in the footnote there. "Beauty is in the eye of the beholder" is the fatuous saying of our time that corresponds to Protagoras's saying. It is obviously true of a certain kind of experience, and is then taken as meaning something altogether different, that there is nothing beautiful in its own right, which we might need clear sight to behold. In the *Nicomachean Ethics,* Aristotle says that bad habits and uncontrolled desires cloud our sight. (1144a 33–36)

moved and everything is moved from something to something, then it is necessary for the thing moved to be in that condition out of which it is going to be moved, and not be in it, and to be moved into this other one and come to be in that, but the things that result in contradiction are not true together at the same times. And if the things around us continually flow and are in motion with respect to how much they are, which one might posit even though it is not true, why should they not stay the same with respect to what sort they are? For contradictory things seem to be attributed to the same thing not least on account of supposing that the sizes of bodies do not remain the same, for which reason the same thing would both be and not be four feet long. But the thinghood of things goes by what sort they are, and this is of a definite nature, while how much they are is indefinite.

1063a 20

Also, why, when ordered by the doctor to take this particular food, do people take it? For why is this a loaf of bread any more than it isn't? So it should make no difference to eat it or not eat it, but as it is, as though these things were true about it and this were the food prescribed, people take it, and yet they ought not to if no nature remains firmly the same among perceptible things, but they are all always in motion and in flux. Also, if we are always changing and never remain the same, what wonder is it if things never seem the same to us, just as happens with those who are sick? (For to them also, since their conditions are not disposed in the same way as when they are healthy, the things that pertain to perceptible things do not appear similar, while the perceptible things themselves do not on this account take part in any change, but make different perceptions, and not the same ones, in those who are sick; and presumably it is necessary that it be the same way also when the change spoken of happens.) But if we do not change but continue being the same, there would be something that stays still.

1063a 30

1063b

So for those who hold on to the impasses mentioned based on reason, it is not easy to refute them if they will not posit something and no longer demand a reason for this, for that is the way every argument and every demonstration comes about, since by positing nothing people annihilate conversation and reason altogether, and so with such people there is no argument, but for those who are at an impasse as a result of the traditional paradoxes, it is easy to meet and to refute the thing producing the impasses in them, and this is clear from what has been said. So it is clear from these things that it is not

1063b 10

possible for contradictory statements to be true about the same thing at one time, nor for contrary statements, since every contrariety is spoken of by way of deprivation, and this is clear to those who reduce the articulations of the contraries to their source. And similarly, neither is it possible to attribute anything that is in between the contraries to one and the same contrary thing, for if the underlying thing is white, then in saying that it is neither black nor white we would be wrong, since it would follow that the same thing was and was not white, since either one of the two intertwined attributes would be true of it, and one of these is the contradictory of white.

1063b 20

So neither is it possible for those speaking in accordance with Heracleitus to be right, nor for those speaking in accordance with Anaxagoras; otherwise contraries would turn out to be attributed to the same thing, for when one says that a part of everything is in everything, he says that nothing is sweet any more than it is bitter, or any of the rest of the contraries whatever, if indeed everything is present in everything not in potency only but at-work and separate. And similarly, not all statements can be false nor all true, both because of many other unmanageable results that could be inferred from positing this, and because if all things were false not even would one be right to say this itself, while if all were true, one would not be wrong to say they were all false.

1063b 30

Chapter 7 Every sort of knowledge seeks certain sources and causes that concern each kind of the things known by them, as do medicine and gymnastic and all the rest of the kinds of knowledge, both productive and mathematical. For each of these, drawing a boundary around some class of things for itself, busies itself about this as something that is and is present to it, but not insofar as it is, but this is the business of another sort of knowledge different from these sorts. Each of the sorts of knowledge mentioned, taking hold in some way of what something is in each class of things, tries to make clear the rest of the things pertaining to it, either more loosely or more precisely. Some of them get hold of what something is by means of perception, others by making a hypothesis, for which reason it is clear from such a survey of examples that there is no demonstration of thinghood or what something is.

1064a

1064a 10

And since there is a kind of knowledge that is about nature, it is clear that it will be different from both the practical and the productive sorts. For in the productive sort, the source of motion is in the one who makes and not in the thing made, and this is either an art or some other capacity, and similarly in the practical sort, the motion is not in the thing done but in the ones who act. But the knowledge of the one who studies nature is about things that have in themselves a source of motion. That, then, it is necessary for the natural sort of knowledge to be neither practical nor productive but contemplative, is clear from these things (since it must fall into some one of these classes). And since it is necessary for each sort of knowledge to know in some way what something is, and to use this as a starting point, one must not let it go unnoticed in what way the one who studies nature needs to define it and how he needs to get hold of the articulation of the thinghood of things—whether in the manner of the snub nose or rather in the manner of the curved line. For one of these, the articulation of the snub nose, is meant as including the material of the thing, but that of the curved line is meant as separate from the material, since snubness comes about in a nose, and hence also the articulation of it is studied along with this, since the snub is a curved-in nose. It is clear then that of flesh and of an eye and of the rest of the parts, one needs always to give the articulation along with the material.

1064a 20

But since there is a sort of knowledge of being as being and as separate, one must examine whether one ought to set this down as being the same as the study of nature or rather as different. Now the study of nature is about things having a source of motion in themselves, while mathematics is contemplative and concerns something that remains the same, but is not separate. Therefore, about the sort of being that is separate and motionless, there is another sort of knowledge that is different from both of these, if there *is* any such independent thing—I mean something separate and motionless—which is just what we shall try to show. And if there is any such nature among beings, that would be where the divine also is, and this would be the primary and most governing source of things. It is clear, then, that there are three classes of contemplative knowledge: physics, mathematics, and theology. So the class of contemplative kinds of knowledge is the best, and among these themselves the best is the one mentioned last, for it is about the most honorable of beings, but each sort of knowledge is called better or worse in accordance with the thing known that is appropriate to it.

1064a 30

1064b

One might be at an impasse whether the knowledge of being as being ought to be set down as universal or not.[11] For each of the mathematical sorts of knowledge is about some one definite class, but universal mathematics is a study that concerns them all in common.

1064b 10   So if natural independent things are primary among beings, then also physics would be the primary sort of knowledge; but if there is another nature and independent thing that is separate and motionless, it is necessary that the knowledge of it be other than and prior to physics, and universal by being prior.

Chapter 8   Since simply to be is meant in more than one way, of which one is meant as incidental being, one must examine first what concerns this sort of being. Now it is clear that none of the kinds of knowledge that have been handed down concerns itself with what is incidental. (For the art of housebuilding does not consider what will

1064b 20   happen incidentally to those who use the house, such as whether they will live there unpleasantly or the opposite, nor do the arts of weaving or leather-working or sauce-making consider such incidental things, but each of these kinds of knowledge considers only what is proper to itself in its own right, and that is the end particular to it; nor about the fact that one who is musical, when he has become literate, will be both of these together, not having been so before, while that which is so, not having always been so, has been coming into being, so that he was coming to be musical-together-with-literate—this is inquired about by none of the kinds of knowledge that are acknowledged to be knowledge, but only by sophistry, for this alone concerns itself with what is incidental, for which reason Plato's assertion was not bad when

1064b 30   he said sophistry spends its time on nonbeing.)

That it is not even possible for there to be knowledge of what is incidental will be clear to those who try to see what in the world it is that is incidental. Now we say of everything either that it is so always and by necessity (meaning necessity not as by force but in the way we use it in matters that have to do with demonstration), or that it is so for the most part, or that it is so neither for the most part nor always and by necessity but however it chances to be, as it might become cold in the dog days, though this happens neither always and by necessity nor

---

[11] On this paragraph, see 1026a 23–32, and footnotes.

for the most part, but might turn out so incidentally sometimes. So the incidental is what happens not always nor by necessity nor for the most part. What, then, the incidental is has been said, and why there is no knowledge of such a thing is clear, for every sort of knowledge is about something that is so always or for the most part, but the incidental is among neither of these.

And that, of what is so incidentally, there are not causes and sources of the same sort as there are of what is so in its own right, is clear, for then everything would be by necessity. For if this is so when that is, and that when this other is, and this last is not however it chances to be but by necessity, then also that of which this is the cause will be by necessity down to the last of the things mentioned as caused (while this was incidental), and so everything will be by necessity, and what can be whichever way it chances and admits of either happening or not is completely annihilated from the things that happen. And if the cause is posited not as being so but as becoming so, the same things will follow, for everything will come to be by necessity. For the eclipse tomorrow will happen if this has happened, and this if something else, and that if another thing, and in this way taking away time from the finite time from now until tomorrow, one will at some time come to something already present, and so, since this is so, all the things that are going to happen after this will come to be by necessity.

So of being in the senses of being true and being incidental, the former is present in something intertwined by thinking and is something undergone within thinking (which is why it is not of being in this sense that the sources are being sought, but of being that is outside and separate), and the latter—I mean what is incidental—is not necessary but indeterminate, and of such a thing the causes are without order and without limit.

That which is for the sake of something is present in things that happen by nature or as a result of thinking, but it is fortune[12] when

---

[12] In the *Physics*, Bk. II, Ch. 4–6, from which this paragraph is taken, Aristotle is careful to distinguish chance in general from fortune, which is chance that occurs within events following from human choices. Here the distinction is muddled by the compression of the argument. In the *Physics* the emphasis is on the fact that ends or final causes are present in all natural events, even though thinking and choice are not. It is the more general conclusion that is taken up here, that chance causes always come about by the interference of two or more lines of causes that are in themselves for the sake of something. By fortune one finds someone who owes him money, when he

1065a 30

any of these happen incidentally, for just as being is in one way in its own right and in another way incidental, so also with cause. And fortune is an incidental cause in the things that are by choice, among those that happen for the sake of something, for which reason fortune and thinking concern the same things, since there is no choice apart from thinking. The causes from which the things that are from fortune might come about are indeterminate, on account of which also fortune is unclear to human reason and is a cause incidentally, but simply it is a cause of nothing. Fortune is good or bad when what issues from it

1065b

is good or bad, but good fortune and ill fortune have to do with great instances of these. And since nothing incidental takes precedence over things in their own right, neither then do incidental causes, so if fortune or chance is a cause of the heavens, intelligence and nature have a prior responsibility.

Chapter 9   Something *is* in one way only as at-work, in another way as in potency, and in another both in potency and at-work, and again in one way as a being, in another as so-much, in other ways in the rest of the categories[13]; and there is no motion apart from things, since something changes always with respect to the categories of being, and there is nothing common to these which is not within a single category.

1065b 10

Each of these belongs to anything in two ways (for example a *this* is either its form or its deprivation, and with respect to the of-what-sort there is white or black, with respect to the so-much there is complete or incomplete, and with respect to change of place there is up or down, or light and heavy), so that there are just as many forms of motion and change as there are of being. Since what is in potency in each particular class of things is distinct from what is at-work-staying-itself, I say that motion is the being-at-work of what is in potency, insofar as it is such. And that we say this truly is clear from the following: for when what is buildable, insofar as we mean it to be such, is at-work, it is being built, and this is the activity of building, and similarly with learning, healing,

---

went to the marketplace to buy oil; by chance the rain that maintains the equilibrium of the cosmos also rots the wheat that was already harvested. The hierarchical ordering of causes is reflected in the distinction between being something in its own right and being something incidentally. See also Bk. VI above, Ch. 2–3.

[13] The "categories" are the eight ways of attributing being listed at 1017a 23–27 (and again below at 1068a 8–10 with one left out), and hence the highest classes.

walking, leaping, growing old, and ripening. It belongs to each to be   1065b 20
in motion whenever this being-at-work-staying-itself is itself present,
and neither before nor after. So the being-at-work-staying-itself of what
is in potency, whenever it is at-work as a being-at-work-staying-itself,
not as itself but as movable, is motion.[14]

I mean the *as* in the following way. Bronze is a statue in potency,
but nevertheless it is not the being-at-work-staying-itself of the bronze
as bronze that is motion. For it is not the same thing to be bronze
and to be potentially-something, since if it were the same simply by its
meaning, the being-at-work-staying-itself of bronze would be a certain
sort of motion. But these are not the same (and this is clear in the case
of contraries, for being capable of being healthy and being capable of
being sick are not the same, since then being healthy and being sick   1065b 30
would be the same, but it is the thing that underlies both being healthy
and being sick, whether this is blood or some other fluid, that is one
and the same), and since they are not the same, just as neither are a
color and being visible the same, the being-at-work-staying-itself of
the potency just as a potency is motion. Now that it is this, and that it
turns out that a thing is in motion at the time when this being-at-work-
staying-itself is itself present, and neither before nor after, is clear. (For   1066a
each thing admits of being at work at one time and at another time
not, such as the buildable as buildable, and the being-at-work of the
buildable insofar as it is buildable is the activity of building, for the
being-at-work is either this, the activity of building, or a house, but
whenever the house is present, it is no longer something buildable,
but what is being built is the buildable thing; therefore it is necessary
that this being-at-work be the activity of building, but the activity of
building is a certain sort of motion, and the same account also applies
to the other motions.)

That this is said rightly is clear from the things other people say
about it, and from the fact that it is not easy to define it otherwise.
For one could not even place it in another general class, and this is   1066a 10
clear from the things people say, for some say motion is otherness

---

[14] What is crucial here is that potencies of things are not just logical possibilities, and
motions are not just brute facts; in a motion a potency itself has the structure of a
being, emerging and holding on as the potency it is by way of activity. See discussions
of this definition in the introduction and commentary of my translation of Aristotle's
Physics, pp. 21–24 and 78–80 (Rutgers University Press, 1995, 1998).

or inequality or nonbeing, none of which is necessarily in motion, but change is no more into or out of these than into or out of their opposites. But the reason for placing it in these classes is that motion seems to be something indefinite, while one whole array of principles seems to be indefinite on account of being negations, since none of them is either a *this* or an of-this-sort or the rest of the categories. And the reason for motion's seeming to be indefinite is that it is not possible to place it as a potency or as a being-at-work of beings, for neither is

1066a 20   what is capable of being so-much necessarily in motion, nor what is actively so-much; and motion seems to be a certain sort of being-at-work, but incomplete, and the reason is that the potency of which it is the [complete] being-at-work is itself incomplete. And for this reason it is hard to grasp what it is, for it is necessary to place it either as a deprivation or as a potency or as an unqualified being-at-work, but none of these seems admissible; so what remains is what has been said, both that it is a being-at-work and that it is the sort of being-at-work that has been described, which is difficult to bring into focus but capable of being.

And it is clear that motion is in the movable thing, for it is the being-at-work-staying-itself of this by the action of the thing capable of causing motion. And the being-at-work of the thing capable of causing motion is not different, since it is necessary that it be the being-at-work-

1066a 30   staying-itself of both; for a thing is capable of causing motion by its potency and is in motion by being-at-work, but it is capable of being-at-work upon the thing moved, so that the being-at-work of both alike is one, just as the interval from one to two and from two to one is the same, and the uphill and downhill road, though the being of them is not one, and similarly also in the case of the thing causing motion and the thing moved.

Chapter 10   The infinite is either that which cannot be gone through because it is not of such a nature as to be gone through, just as sound is invisible, or that which has a way out through it which cannot be completed, or can scarcely be completed, or that which is of such

1066b   a nature as to have, but does not have, a way out through it or a limit, and it is so by way of either addition or subtraction or both. Now it is not possible for it to be itself something separate, for if the infinite were neither a magnitude nor a multitude, but were itself an independent thing and not an attribute, it would be indivisible (since what is divisible is either a magnitude or a multitude); but if it is indivisible it is not infinite except in the way that sound is invisible, but this is not what people mean by it nor what we are inquiring about, which is that which has no way out through it. Also, how could the infinite admit of being something in its own right, if number and magnitude, of which the infinite is an attribute, do not? Also, if it is an attribute, it could not, insofar as it is infinite, be an element of beings, any more than the invisible is an element of speech, despite the fact that sound is invisible. And it is clear that it is not possible for there to be an infinite actively. For then any part of it whatever that was taken would be infinite (since the infinite and the being-infinite of it are the same, if the infinite is an independent thing and does not belong to some underlying thing), and so it is either indivisible or, if divisible, divided into infinites; but it is impossible for the same thing to be many infinites (for just as a part of air is air, so would a part of the infinite be infinite, if it is an independent thing and a source), and therefore it is indivisible and without parts. But this is impossible for what is infinite in complete activity (since it is necessary that it be a so-much), and therefore it is present as an attribute. But if it has being in that way, it has been said that it cannot be *it* that is a source of things, but that of which it is an attribute, the air or the even.[15]

    This inquiry has pertained to universals, but that there is no infinite among perceptible things is clear from what follows. For if the articulation of a body is that which is bounded by surfaces, there could not be an infinite body, neither a perceptible one nor an intelligible solid; nor could there be a number that is separate and infinite, since a number or that which has a number is countable. And this is also clear from the following considerations that pertain to the study of nature. For the infinite could not be either composite or simple. It could not be a composite body, if the elements are finite in number (for it is necessary that the contrary elements be in equilibrium, and that no one of them is infinite, since if the power in either of the bodies falls short to any degree whatever, the finite one will be destroyed by the infinite one— and it is impossible that each of them be infinite, since a body is that

1066b 10

1066b 20

1066b 30

---

[15] Anaximenes said that the source of all things was the infinite air. For the Pythagoreans, who said everything is number, the odd was the source of finitude and identity, while the even, that breaks in two, was the source of the infinite.

which has extension in every direction, and the infinite is what is extended limitlessly, so that if there is an infinite body, it will be infinite in every direction); but neither is it possible for an infinite body to be one and simple, neither as some people say, as something apart from the elements, out of which they are generated (for there is no such body apart from the elements, since everything is made of something and dissolves into this, but this is not apparent with anything besides the simple bodies), nor as fire or any other of the elements, for apart from the being-infinite of any one of them, it is impossible for the sum of things, even if it is finite, either to be or to become one of them, as Heracleitus says that all things sometimes become fire. The same argument also applies to the one body that the writers on nature make besides the elements, for everything changes out of a contrary, as out of hot into cold.

1067a

Also, a perceptible body is somewhere, and the same place belongs to the whole and to the part, as with the earth, so if it is homogeneous, it will be motionless or always carried along, but this is impossible. (For why downward rather than upward or in any direction whatever? For instance, if it were a lump of earth, where would this be moved or where would it stay still? For the place of the body homogeneous with it is infinite. Then will it take up the whole place? And how? What then will its rest or its motion be? Or is it at rest everywhere? Then it will not be moved. Or will it be in motion everywhere? Then it will not stay still.) But if the whole is heterogeneous, then the places are also heterogeneous, and, first, the body of the whole will not be one other than by contact; further, the parts will be either finite or infinite in kind. It is not possible for them to be finite (for then some, such as fire or water, will be infinite in extent and others not, if the whole is infinite, but such things would be the destruction of their contraries), and if they are infinitely many and simple, and the places too are infinite, then the elements will also be infinite; but if this is impossible and the places are finite, then the whole will also necessarily be bounded.

1067a 10

1067a 20

In general it is impossible for there to be an infinite body and a place for bodies if every perceptible body has either heaviness or lightness; for a body will be carried either to the center or upward, but it is impossible for the infinite, either the whole or the half of it, to be affected in either of these ways. For how would you cut it in two? Or how, in the infinite, will there be up and down, or extremity and center? Also, every perceptible body is in a place, and of place there

are six forms,[16] but it is impossible for these to be in the infinite body. 1067a 30
And in general, if it is impossible for a place to be infinite, then it is also
impossible for a body, for what is in a place is somewhere, and this
means either up or down or any of the rest, and each of these is some
kind of limit. And the infinite is not the same thing in magnitude, in
motion, and in time, as though it were some one nature, but the one
that is derivative is called infinite as a consequence of the one that takes
precedence, as a motion is called so as a consequence of the magnitude
along which it moves or alters or grows, and the time on account of
the motion.

Chapter 11   One sort of thing that changes does so incidentally, in 1067b
the sense that an educated person walks, and another sort is said to
change simply, on account of the changing of something that belongs
to it, such as those that change on account of their parts (for the body
gets well because the eye does), but there is something that is moved
primarily on account of itself, and this is what is moved in its own right.
And this is the same way also with the thing that causes motion, for it
does so either incidentally, or on account of a part, or in its own right,
and there is something that primarily causes motion, and something
that is moved, and also the time in which it is moved and that from
which and to which it is moved. But the forms and the attributes and 1067b 10
the place, to which the moving things are moved, such as knowledge
and heat, are motionless; it is not heat that is a motion but the process
of heating. Change that is not incidental is not present in all things but
in contraries and what is between them and in contradictories, and
belief in this comes from considering examples.
   A thing that changes does so either from one underlying thing to
another, or from what is not a subject to what is not another subject, or
from a subject to what is not that subject, or from what is not a subject
to that subject (and by "subject" I mean what is declared affirmatively),
so that there must be three kinds of change, since that from what is
not one subject to what is not another subject is not a change, for they 1067b 20
are neither contraries nor is there a contradiction, because there is no
opposition between them. That which is from what is not a subject to a
contradictory subject is coming-into-being, simply coming-into-being

---

[16] These are up and down, before and behind, right and left.

when the change is simply into that subject, or particular becoming when it is a change of some particular attribute; that from a subject to what is not that subject is destruction, simple passing away when the change is simply out of that subject, or particular destruction when it is out of something particular.

Now if what-is-not is meant in more than one way, and what is not so because of [faulty] combination or separation [in speech or thought] does not admit of being moved, nor does what is by way of potency, which is opposed to what *is* simply (for the not-white or not-good still admit of being moved incidentally, since the not-white could be a human being, but what is simply not a *this* in no way admits of it), then it is impossible for what-is-not to be moved. (And if this is so, then it is also impossible for coming-into-being to be a motion, for what passes into being is what is not. For however much it comes to be incidentally, it is still true to say that not-being belongs to what simply comes into being.) And it is likewise impossible that what-is-not be at rest. These awkward consequences also follow if everything that is moved is in a place, since what-is-not is not in a place, for it would then *be* somewhere. And destruction is not a motion either, for the opposite of a motion is either a motion or rest but destruction is the opposite of coming-into-being. And since every motion is a change, and the kinds of change mentioned are three, but of these those that result from coming-into-being or destruction are not motions, and these are the changes that are between contradictories, it is necessary that change from one subject to another be the only sort of change that is motion. And the underlying subjects are either contraries or in-between (for let a deprivation also be set down as a contrary), and are declared affirmatively, such as naked, baregummed, and black.[17]

Chapter 12 So if the ways of attributing being are divided into thinghood, quality, place, acting or being acted upon, relation, and quantity,[18] there are necessarily three kinds of motion, with respect to

[17] A motion is always a continuous transition from one positive condition, through in-between states, to its contrary, and hence is always rooted in what a thing is.

[18] The eighth category, time, is omitted because it is not the end point of any motion but that in which they all take place. See 1067b 9–10. This listing is also untypical in that most of the categories are expressed as general nouns (such as place) rather than as interrogative terms made into substantives (such as the where). One should

1067b 30

1068a

1068a 10

the of-what-sort, the how-much, and the place. There is no motion with respect to thinghood, because nothing is contrary to an independent thing, nor of relation (since it is possible, when one of the two related things changes, for the relation not to be true, even though the other thing has not changed in any way, so that the motion of them is incidental), nor is there a motion of acting and being acted upon, nor of moving and being moved, because there is not a motion of a motion or a coming into being of coming into being, or generally a change of a change. For there could be two ways of there being a motion of a motion, either in the sense that it is a motion of an underlying motion (for example, a human being is in motion because he is changing from pale to swarthy, so in that way too a motion is either heated or cooled or alters its place or grows; but this is impossible, since change is not any of the things underlying change), or by way of some other underlying thing's changing out of one change into a change of another form, as a human being changes from sickness into health. But this is not possible either, except incidentally. For every motion is a change from one thing to another, and this is so also with coming into being and destruction, except that these are changes into one sort of opposites, while motion is a change into another sort. So at the same time someone is changing from health into sickness, and from this change itself into another. But it is clear that when he has become sick, he will have changed into whatever condition it is (for he could have come to rest), and further that this is not always into whatever happens along. And the other change will be from something into something else, and so it will be the opposite change, getting well; otherwise it will be by having been incidental, as there is a change from remembering to forgetting because that to which they belong changes, at one time into knowledge, at another into ignorance.

1068a 20

1068a 30

Also, it would go to infinity if there were to be a change of a change and a coming into being of coming into being. And an earlier one would be necessary if a later one were to be; for example, if a simple coming into being at some time came into being, and the coming into being of it came into being, so that not yet would there be the thing that came

1068b

not read too much into these discrepancies, but look to the immediate context. (In a somewhat similar way, in the last paragraph of Ch. 8 above, the same word is used for chance in general and fortune in particular. Book XI adapts to its purposes arguments that originate elsewhere, without rewriting them.)

into being simply, but a coming into being that was coming into being beforehand, and this in turn came to be at some time, so that not yet would there be the thing that was coming into being then. And since of infinite things there is no first one, there would not be a first becoming, and therefore no next one either, and then nothing could either come into being or be moved or change. Further, to the same thing there belongs a contrary motion and a state of rest, and both a coming into being and a destruction; therefore what comes into being, whenever it comes into coming-into-being, is at that time being destroyed, for neither at the outset, nor after it has come to be, is it a thing coming into being, and what is being destroyed must *be*. Also, it is necessary for material to underlie what becomes and what changes. What then would it be? Just as body or soul is the thing that is altered, what in that way is the thing that becomes motion or becoming? And again, what is that toward which they are moved? For motion or becoming must be of something, from something, to something. But how? For there could be no learning of learning, so neither could there be a coming into being of coming into being.

1068b 10

And since there is no motion of thinghood or of relation or of acting and being acted upon, it remains that there is motion with respect to the of-what-sort, the how-much, and the place (for contrariety belongs to each of these), and by the of-what-sort I mean not what is in the thinghood of a thing (since then even the specific difference would be a quality), but something to which it is passive, by which a thing is said to be either affected or unaffected.[19] And what is motionless is either that which is altogether incapable of being moved, or that which is scarcely moved in a long time or begins slowly, or what is of such a nature as to be moved and is capable of it, but is not moved at some time when or place where and manner in which it is natural to it, which is the only one of the motionless things that I speak of as at rest, for rest is contrary to motion, and so would be a deprivation of motion in that which admits of it.

1068b 20

Coincident in place are those things that are in one primary place, and separate are those that are in different ones; touching are those

---

[19] The fact that there is qualitative change does not mean that every qualitative attribute of a thing is changeable. What belongs to its thinghood is part of what actively maintains it as what it is; what changes and can be absent is only passively present.

things of which the extremities are coincident, and the in-between is that at which a changing thing naturally arrives before it changes to what is by nature last, when it changes continuously. What is contrary with respect to place is what is as far away as possible along a straight line; next in series is that which, being after the beginning in position or in form or in some other determinate way, has nothing of the same kind between it and that to which it is next in series, such as lines if it is a line or units if it is a unit or houses if it is a house (but nothing prevents something else from being in-between). And what is next in series is next in series to something, and is something that follows, for one is not next in series to two, nor the first day of the month to the second. That which, being next in series to something, is touching it, is next to it.

1068b 30

1069a

Now since every change is between opposites, and these are contraries or contradictories, but there is nothing between contradictories, it is clear that what is in-between is between contraries. The continuous is the same as what is next to something, but I call things continuous when the limits of each of them at which they are touching and holding together have become one and the same, and so it is clear that the continuous is among those things out of which some one thing naturally comes into being as a result of the uniting. And it is clear too that the first of these relations is next in series (for what is next in series is not necessarily touching, but what is touching is next in series, and if things are continuous, they are touching, but if they are touching they are not on that account continuous, and in those things in which there is no contact, there is no growing into one). And so it is not possible for a point and a numerical unit to be the same, since it belongs to points to be [coincident],[20] but to units not even to be touching, but to be next in series, and with distinct points there is something in-between, but with units there is not.

1069a 10

---

[20] Both here and at *Physics* 227a 29 Aristotle has the word "touching," but his meaning is clear enough. These last two paragraphs have the effect of displaying that the variable conditions of changing things belong to orderings as precise as the successions in arithmetic and the continuities in geometry.

Book XII (Book Λ)
The Cause of Being[1]

Chapter 1 Our study concerns thinghood, for it is the sources and 1069a 18 causes of independent things that are being sought. For if the sum of things is some kind of whole, thinghood is its primary part, but 1069a 20 also if the sum of things is just one thing next to another, even so the thinghood of anything is primary, of-what-sort it is comes next, and then how much it is. And at the same time, these latter are not even beings, if one speaks strictly, but qualities and motions, or else even the not-white or not-straight would be beings; certainly we say that even these things have being, for example that the not-white *is*. What's more, none of the others is separate. And the ancient thinkers bear witness to this by what they *do*, since they are looking for the sources and elements and causes of thinghood. Accordingly, while people now assume that it is rather universals that are independent things (for general classes are universal, and it is these that they say are more so sources and independent things, because they inquire from a logical standpoint), those in early times assumed that the independent things were particulars, such as fire and earth, and not the common 1069a 30 class, body.

And there are three kinds of thinghood: One kind is perceptible, of which one sort is everlasting and the other sort, which everyone acknowledges, is destructible, as with plants and animals, and it is necessary to grasp the elements of this kind of thinghood, whether there is one or many. Another kind is motionless, and some people say that this is separate, some of them dividing it into two kinds, others making the forms and the mathematical things into one nature, and still others assume only the mathematical things among these. And the former kinds belong to the study of nature (since they include motion), 1069b but this kind belongs to a different study, if there is no source common to them.

Now perceptible thinghood is changeable, and if change is from opposites, or from things between opposites, but not from all kinds of opposites (since a sound is not-white), but only from a contrary, it is

---

[1] This title for Book XII supplied by the translator.

necessary that there be something underlying the thing that changes into the contrary condition, since the contraries themselves do not change.

Chapter 2 Further, there is something that persists, while the contrary state does not persist; therefore there is a third thing besides the contraries, the material. So if there are four kinds of changes, with re-

1069b 10   spect to what something is, of-what-sort it is, how much it is, or where it is, and simple coming-into-being or passing-away is change of the *this*, increase or decrease is change of the how-much, alteration is change of an attribute, and local motion is change of place, changes are into contrary states of these particular kinds. So the material has to be capable of changing into both, and since being is of two kinds, everything changes from something that has being in potency to something that has being at-work (as from potentially white to something actively white, and similarly too in the case of increase and decrease), so that things are able not only to come into being from what, in an incidental sense, is

1069b 20   not, but also everything comes into being from what is, though from what is potentially but is not at work. And this is Anaxagoras's "one," since it is better than "everything together," and it is also the "mixture" spoken of by Empedocles and Anaximander, and what Democritus is talking about, but for us it is "all things were in potency, but not at-work." Therefore, these thinkers did get some grasp of material, but all things have material, all those that change, but different sorts of material; and material belongs to those everlasting things that are not generated but moved by change of place, though it is not material for coming into being but for from-somewhere-to-somewhere. (And one might in fact be at an impasse about what sort of not-being generation comes out of, since not-being is of three kinds.[2])

Now if something has being in potency, still this is not a potency to be any random thing, but a different thing comes to be from a different potency, and it is not sufficient to say that [even as potencies] all things

1069b 30   are there together, for they differ in their material; why did they come to be an infinity of things and not just one? For intellect is one, so that if material also were one, that which the material was in potency would

---

[2] See the beginning of Bk. IX, Ch. 10.

have come to be actively.[3] So the causes are three, as are the sources, two of them being the pair of contraries of which one is the articulation or form and the other its deprivation, while the third is the material.

Chapter 3 Next after that is the fact that neither material nor form comes into being—I am speaking of the ultimate ones. For with everything, something changes, by the action of something, and into   1070a
something: that by the action of which it changes is the first thing that sets it in motion, that which changes is the material, and that into which it changes is the form. So this is an infinite process if it is not only the bronze that comes to be round, but the roundness or bronze too comes into being; and so it necessarily comes to a stop. And next after this is the fact that each independent thing comes into being from something that has the same name (for this is true both of natural independent things and of the rest). For it comes into being either by art or by nature, or else by fortune or chance. Now art is a source that is in something else, but nature is a source that is in the thing itself (since a human being begets a human being), while the rest of the causes are deprivations of these.[4]

The kinds of thinghood are three, since the material is a *this* by   1070a 10
coming forth into appearance (for whatever has being by way of contact, and not by having grown together, is material and underlies something else), while the nature of a thing is a *this* and an active condition into which it comes; and then the third kind is the particular thing that consists of these, such as Socrates or Callias. Now in some cases the *this* is not present apart from the composite independent thing, as the form of a house does not have being apart, unless the art does (though there is no coming-into-being or destruction of these forms, but it is in a different way that a house without material, and

---

[3] These are references to the claims of the thinkers named earlier in the chapter, whose complicated theories are replaced by Aristotle's own claim in the next sentence. The chapter is a condensed version of the main argument in *Physics* Book I.

[4] This is a sketchy summary of things argued at length elsewhere. In nature, the *same* form that is at work in the parent comes to be present in the offspring; in art, the form that is at work upon the artisan's intellect and imagination is imposed from without upon material; fortune is the achievement of some human purpose by an accidental combination of other causes, and chance is the achievement of any end whatever, including natural ones, in that way. See Bk. VII, Ch. 7–9, above, and *Physics* Book II.

health, and everything that results from art is and is not[5]), and if it is present apart, it is in the case of things that are by nature; and so for this reason Plato said not badly that the forms are as many things as there are by nature,[6] if there are forms, though there are none of things

1070a 20   like fire, flesh, or a head, since all these are material, and the final sort of material belongs to what is an independent thing most of all.

Now things that cause motion are causes as being previously present, but things that are causes in the sense of rational patterns are simultaneous with what they produce. For at the time when a human being is healthy, then too the health is present, and the shape of a bronze sphere is present at the same time as the bronze sphere. But one must examine whether any of these also remain afterward; for nothing prevents it in some cases, for instance if the soul is of this sort, not all of it but the intellect, since presumably this is impossible for the whole soul. So it is clear that there is no need on account of *these* reasons for there to be forms, since a human being begets a human being, a particular one begetting a particular one, and similarly too

1070a 30   with the arts, since the medical art is the reasoned account of health.[7]

Chapter 4 Now there is a sense in which the causes and sources of different things are different, but there is a sense in which, if one speaks universally by way of analogy, they are the same for all things. For one might be at an impasse whether different sources and elements, or the same ones, belong to independent things and to relations, and similarly with each of the different ways of attributing being. But it is absurd that they should be the same for everything, for then relations

1070b   and thinghood would come from the same things. What would these

---

[5] Aristotle does not think that the artist or artisan ever creates or invents any form. The doctor must grasp health as the potency of human bodies, and the architect must discover the possibilities in materials to serve as shelters. Previously unnoticed possibilities of things were in them, and present among objects of thought and imagination, all along.

[6] Whether this is said in the dialogues, and if so where, is a question one might enjoy pursuing. The "couch in nature" of *Republic* 597B might be mentioned; if it is the same as the bed of rushes at 372B, it gives an example of how an artisan can bestow new products on the world, without creating any new form.

[7] The necessity that there be forms that are separate and at work does not come out of this general analysis of becoming, but arises out of answering the question raised in Bk. VII, Ch. 17: What makes anything one if it is not a heap but a genuine whole?

sources be? For aside from thinghood and the other ways of attributing being, there is no one thing common to them, but an element precedes those things of which it is an element. But surely thinghood is not an element of relations, nor is any relation an element of thinghood. What's more, in what way could the same elements belong to all things? For none of the elements could be the same as that which is composed of elements, as B or A is not the same as BA (and so none of the intelligible things, such as being or oneness, is an element, since they belong also to each one of the composite things). Therefore none of the elements could be either an independent thing or a relative thing, but they would have to be of one sort or the other. Therefore, the same elements cannot belong to all things. Or else, as we are saying, in one sense they do, but in another sense they do not; for example, of perceptible bodies, perhaps the source in the sense of form is the hot, and in another way the cold, as its lack, while the material is the first thing that is in virtue of itself potentially these, and these are the thinghood of perceptible bodies, as is also what consists of them, of which they are the sources, or anything consisting of the hot and the cold that might come to be one, such as flesh or bone, since what comes from different things is different.

1070b 10

Accordingly, these perceptible things have the same elements and sources (though different things have different ones of them), but it is not possible to say in this way that *all* things have the same elements and sources, except by analogy, just as if one were to say that there are three kinds of sources: form, deprivation, and material. But each of these is different as it concerns each class of things; for instance, among colors they are white, black, and a surface, or light, darkness, and air, out of which come day and night. But since it is not only the constituents of things that are their causes, but these may also be found among external things, such as that which sets them in motion, it is clear that a source and an element are different, though both are causes, and sources are divided into these kinds, the source that sets something in motion or brings it to a stop being one of them, so that there are by analogy three kinds of elements, but the causes and sources are of four kinds[8]; but the elements are different in different things, and the first

1070b 20

8 This does not refer to the four causes of *Physics* Book II, but just totals up the kinds mentioned here. The external source of motion is understood as a cause of generation,

cause that sets them in motion is also different in different things. With health, disease, and a body, the mover is the medical art. With form, a certain particular kind of disorder, and bricks, the mover is the house-building art. And since among natural things the mover is, say, for a human being a human being, while among things that are produced by thinking it is the form or its contrary, the causes would again in a certain way be three, though by this means four. For health is in a certain way the medical art, and the form of a house is the builder's art, and a human being begets a human being; but still, over and above these, is the cause which, as the first of all things, sets all things in motion.

Chapter 5 Now since some things are separate while others are not separate, the former are independent things.[9] And it is on account of this that all things have the same causes, because without independent things, attributes and motions are not possible. So then these causes will be, presumably, soul and body, or intellect, desire, and body. And in yet another way, the sources of things are the same by analogy, namely being-at-work and potency, though these are both different and present in different ways in different things. For with some things, the same thing is present at a certain time actively which was at another time in potency, such as wine or flesh or a human being. (And these also fall into the kinds of causes mentioned, since the form is at work if it is separate, as is the composite of both form and material, and even a deprivation such as darkness or [a composite of a deprivation and material] such as a sick person, while the material has being as potency, since it is what is capable of becoming both the form and its deprivation.) But being-at-work and being in potency differ in different ways in things in which there is not the same material, for which there is not the same form but a different one, just as the causes of a human

1070b 30

1071a

1071a 10

---

that by which the material is moved from lacking to having the form that will make the thing what it is. The final cause is either identical to, or included in, that form.

[9] Separateness is the primary mark of anything that has thinghood. It does not refer to separateness *from* anything or any place, but to anything that is intact on its own. It is a synonym for being a *this,* anything that presents itself to us as a distinct item—first of all, for us, in sensory experience, but also in the widest sense. The idea of wholeness, as belonging to something that is more than a sum of parts, arises from reflection, but separateness is what we get hold of first and directly. It is consistently the beginning of Aristotle's thinking about thinghood.

being are elements, fire and earth as material and the proper form, and in addition something else outside, such as a father, and besides these the sun and its slanted orbit,[10] which are not material or form or deprivation or of the same kind, but causes of motion.

Further, it is necessary to see that some things are possible to state universally, but others not. Now the primary sources of all things are the *this* that is first at work and something else which is in potency. So these are not the universal causes, since the source of particular things is particular; for a human being is the source of a human being universally, but no one is this universal, but rather Peleus is the source of Achilles and your father of you, and this particular B is the source of this particular BA, but B in general is the source of BA simply. And then, if the causes and elements of independent things are the sources of all things (but different ones of different ones), then as was said, of things not in the same class (colors and sounds, or independent things and quantity) they are different except by analogy; of things that are in the same kind they are also different, but not in kind, but because they are different for particular things, your material and form[11] and mover from mine, though they are the same in their universal statement.

So as for seeking out what are the sources or elements of independent things and of relations and the of-what-sorts of things, and whether they are the same or different, it is clear that, since they are meant in more than one way, they do belong to everything, but when they have been distinguished they are not the same but different, except in one sense. And the causes of all things are the same in this sense—by analogy—because they are material, form, deprivation, and a mover, and the causes of independent things are the causes of all things in this sense—because when they are taken away everything is taken away; and further, the primary thing that is completely at work

1071a 20

1071a 30

---

[10] It is the cyclical departure and approach, at any place on the earth, of the sun on the ecliptic, that causes the seasons of growth and renewal which provide one necessary condition of human or any other life, namely food. The sun is not at one time active, at another potential, but always in motion, with a being-at-work *of* a potency to be always approaching and departing from every part of the earth.

[11] This is the only place in which Aristotle speaks of forms as particular in this sense, as individual souls. The sense in which he consistently regards forms as particular becomes explicit in the next two chapters; they are everlasting particulars in the activity of the divine intellect. It is in this sense that each form is separate and a *this* (1017b 27–28), which could not be said of an individual soul.

1071b

is the cause of all things. But the causes are different in this sense—they are as many as there are primary contraries, described neither generically nor ambiguously, and as there are kinds of material as well. So what the sources are of perceptible things, and how many there are, and in what way they are the same and in what way different, have been said.

Chapter 6 Now since there are three kinds of thinghood, two of them natural and one motionless, about the latter one must explain that it is necessary for there to be some everlasting motionless independent thing. For independent things are primary among beings, and if they were all destructible, everything would be destructible; but it is impossible for motion either to come into being or to be destroyed (since it always is), and impossible too for time. For if there were no time, there could be no before and after; and motion is continuous 1071b 10 in just the way that time is, since time is either the same as or some attribute of motion. But there is no continuous motion other than in place, and among these, other than in a circle.

But surely if there is something capable of moving and producing things, but not at work in any way, there will not be motion; for what has a potency admits of not being at work. Therefore, there is no benefit even if we adopt everlasting independent things, as do those who bring in the forms, unless there is in them some source capable of producing change; moreover, even this is not enough, not even if there is another independent thing besides the forms, since if it is not going to be at work, there will not be motion. What's more, it is not enough even if it will be at work, if the thinghood of it is potency, for there would not be everlasting motion, since what has being in potency admits of not 1071b 20 being. Therefore it is necessary that there be a source of such a kind that the thinghood of it is being-at-work. On top of that, it is necessary that these independent things be without material, for they must be everlasting, if indeed anything else is everlasting. Therefore they are being-at-work.[12]

---

[12] The stars themselves and their motions are considered to be everlasting, despite the presence of material, but this relies on some cause that is unceasingly at work upon

And yet there is an impasse: for it seems that, while everything that is at work is capable of it, not everything that is capable of it is at work, so that the potency would take precedence. But surely if this were so, there would be no beings at all, since it is possible to be capable of being and not yet be. Nevertheless, there is the same impossibility if things are the way those who write about the gods say, who generate all things out of night, or the way of those who write about nature, who say "all things were together." For how will things have been set in motion, if there were not some responsible thing at work? For material itself, at any rate, will not set itself in motion, but a craftsman will   1071b 30 cause it to, nor will the menstrual fluid or the earth set themselves in motion, but semen or seeds will cause them to. And this is why some people, such as Leucippus and Plato, bring in an everlasting activity, for they say that there is always motion. But why there is this motion, and what it is, they do not say, nor the cause of its being a certain way or some other way. For nothing moves at random, but always something must be present to it, just as now something moves in a certain way by nature, but in some other way by force or by the action of intelligence or something else. And then, what sort of motion is primary? For this makes so much difference one can hardly conceive it. But surely it is not even possible for Plato to say what he sometimes thinks the source   1072a of motion is, which itself sets itself in motion; for the soul is derivative, and on the same level as the heavens, as he says.[13]

Now to suppose that potency takes precedence over being-at-work is in a sense right but in a sense not right (and in what sense has been said); and Anaxagoras testifies that being-at-work takes precedence (since intellect is a being-at-work), as does Empedocles with love and strife, and so do those who say there is always motion, such as Leucippus; therefore there was not chaos or night for an infinite

---

them. At the end of the line there must be something in which there is no possibility of being otherwise, the very being of which is to be at work.

[13] This is a reference to the *Timaeus,* though elsewhere (especially *Phaedrus* 245C–46A), Socrates speaks of the soul as the ultimate source of all motion. The speakers differ in these dialogues, the *Timaeus* deliberately depicts eternal things as coming to be in time, and Socrates habitually makes dialectical arguments that are faulty if taken out of context, but Aristotle's point is simply that we cannot get any help on this question from Plato's writings.

time, but the same things have always been so, either in a cycle or in some other way, if being-at-work takes precedence over potency. So

1072a 10   if the same thing is always so in a cycle, it is necessary for something to persist always at work in the same way. But if there is going to be generation and destruction, there must be something else that is always at work in different ways. Therefore it must necessarily be at work in a certain way in virtue of itself, and in another way in virtue of something else, in virtue, that is, of either a different thing or the first one. And it is necessarily in virtue of the first one, since it would in turn be responsible for both itself and that different one. Accordingly it is better that it be the first one, for it was responsible for what is always the same way, while another thing was responsible for what happens in different ways, and obviously both together are responsible for what happens in different ways always. And without doubt motions are this way. Why then must one look for other sources?

Chapter 7 But since it is possible for it to be this way, and if it is not

1072a 20   this way things will come from night and from "all things together" and from not-being, these questions could be resolved; and there is a certain ceaseless motion that is always moving, and it is in a circle (and this is evident not only to reason but in fact), so that the first heaven would be everlasting. Accordingly, there is also something that moves it. And since what is in motion and causes motion is something intermediate, there is also something that causes motion without being in motion, which is everlasting, an independent thing, and a being-at-work. But what is desired and what is thought cause motion in that way: not being in motion, they cause motion. But the primary instances of these are the same things, for what is yearned for is what seems beautiful, while what is wished for primarily is what is beautiful[14];

---

[14] Aristotle uses the word for the beautiful rather than the good here. We wish for all sorts of intermediate and instrumental goods, but ultimately, and therefore primarily, for what they are all for the sake of, which is in turn only for its own sake (see 994b 9–15). In the *Nicomachean Ethics,* Aristotle divides all goods into the beneficial, the pleasant, and the beautiful (1104b 31), and identifies the beautiful as the aim of all moral virtue (1115b 13). This is not an "aesthetic" sense of beauty, which would just be one kind of pleasure, but what we mean when we speak of something as a beautiful thing to do, one in which everything is right.

but we desire something because of the way it seems, rather than it's
seeming so because we desire it, for the act of thinking is the beginning.   1072a 30
But the power of thinking is set in motion by the action of the thing
thought, and what is thought in its own right belongs to an array of
affirmative objects[15] of which thinghood is primary, and of this the
primary kind is that which is simple and at work. (But what is one
and what is simple are not the same, for oneness indicates a measure,
but what is simple is itself a certain way.) But surely the beautiful
and what is chosen in virtue of itself are also in that same array, and
what is primary is always best, or analogous to it. And that-for-the-   1072b
sake-of-which is possible among motionless things, as the [following]
distinction makes evident; for that-for-the-sake-of-which is either *for*
something or *belonging to* something, of which the former is and the
latter is not present among motionless things.[16] And it causes motion
in the manner of something loved, and by means of what is moved
moves other things.

   Now if something is moved, it admits of being otherwise than it is;
and so, even if the primary kind of change of place is a being-at-work,
insofar as something is moved, it is in that respect at least capable
of being otherwise, with respect to place even if not with respect to
thinghood. But since there is something that causes motion while being
itself motionless, this does not admit of being otherwise than it is in
any respect at all. For among changes, the primary one is change of
place, and of this the primary kind is in a circle, but this is what this   1072b 10
mover causes. Therefore it is something that has being necessarily, and
inasmuch as it is by necessity it is beautiful and in that way a source.
For the necessary has this many senses: what is by force because it is
contrary to a thing's impulse, that without which something will not
be in a good condition, and that which does not admit of being any

---

[15] See 1004b 27–1005a 12.

[16] The latter sense is the normal one, in which the things that move and act do so for
the sake of their own self-maintenance, and could not apply to anything motionless.
The former sense is distinguished at *Physics* 194a 35–36, to show that the human
appropriation of other living things to our use does not make them cease to be for
their own sakes, and at *On the Soul* 415b 2–3, to expand the self of each living thing
to include its offspring. Here it is the cosmos, the visible heaven, that is for its own
sake but can also be for the sake of something motionless.

way other than in a simple condition.[17] On such a source, therefore, the cosmos and nature depend.[18]

And the course of its life is of such a kind as the best we have for a short time. This is because it is always the same way (which for us is impossible), and because its being-at-work is also pleasure (which is what makes being awake, perceiving, and thinking the most pleasant things, while hopes and memories are pleasant on account of these). And the thinking that is just thinking by itself is a thinking of what is best just as itself, and especially so with what is so most of

1072b 20   all. But by partaking in what it thinks, the intellect thinks itself, for it becomes what it thinks by touching and contemplating it, so that the intellect and what it thinks are the same thing. For what is receptive of the intelligible and of thinghood is the intellect, and it is at work when it has them; therefore it is the being-at-work rather than the receptivity the intellect has that seems godlike, and its contemplation is pleasantest and best. So if the divine being is always in this good condition that we are sometimes in, that is to be wondered at; and if it is in it to a greater degree than we are, that is to be wondered at still more. And that is the way it is. But life belongs to it too, for the being-at-work of intellect is life, and that being *is* being-at-work, and its being-at-work is in itself the best life and is everlasting. And we say that it is a god who everlastingly lives the best life, so that life and

1072b 30   continuous and everlasting duration belong to a god; for this being is god.

And those who assume, as do the Pythagoreans and Speusippus, that what is most beautiful and best is not present in the source of anything, since, while the sources of plants and of animals are responsible for them, what is beautiful and complete is in the effects

---

[17] Book VIII of the *Physics* traces all change to circular change of place. As its cause, that which is being described here is outside the realm of change and is what *is* simply. That way of being is called in Bk. V, Ch. 5, above, the primary kind of necessity. It is not the necessary means to what is good, but the precondition for there being something variable that can have a good condition. It is called beautiful as the stable simplicity that makes order possible.

[18] The cosmos depends on it as its motionless source of motion, but nature depends on it as the source of unity in everything that is whole, namely the animals and plants. It is only an object of thought that can cause motion without being moved, but it is only an act of thinking that can cause the highest kind of unity (1052a 29–33). The following paragraph shows how the same being is both thinker and thought, and thus the cause at once of all being and all motion.

that come from them, do not think rightly. For the seed comes from other, earlier, complete beings, and what is first is not the seed but the complete being, just as one would say that a human being precedes the germinal fluid, not the one who comes into being from it, but another one from whom the germinal fluid came.

1073a

That, then, there is an independent thing that is everlasting, motionless, and separate from perceptible things, is clear from what has been said. And it has also been demonstrated that this independent thing can have no magnitude, but is without parts and indivisible (for it causes motion for an infinite time, while no finite thing has an infinite power, and since every magnitude must be either infinite or finite, it cannot have magnitude, either finite, for the reason given, or infinite, because there is no infinite magnitude at all[19]). But surely it has also been demonstrated that it cannot be affected or altered, since all other motions are derivative from change of place.[20] So it is clear why these things are this way.

1073a 10

Chapter 8 But whether one must set down one or more than one such independent thing, and how many, must not go unnoticed; but one has to mention, as far as the pronouncements of others are concerned, that about the number of them they have said nothing which can even be stated clearly. For the assumption about the forms contains no particular speculation about it (for those who speak of the forms say the forms are numbers, and about the numbers, sometimes they speak as though about infinitely many, but sometimes as though they had a limit at the number ten,[21] but why the multitude of the numbers is just so much, nothing is said with a serious effort at demonstrative reasoning). But it is necessary for us to argue from the things that have been laid down and distinguished. For the source and

1073a 20

---

[19] This is argued at greater length at the very end of the *Physics,* drawing on the whole content of that work. This sentence, in effect, incorporates the whole of the *Physics* into the conclusion of the *Metaphysics.*

[20] The source of change of place, since it has no magnitude, has no place, and consequently could not change in that way or any other.

[21] This is attributed to Plato at *Physics* 206b 32. It is eidetic numbers, assemblages of forms with a structure analogous to that of numbers, that are meant. See 987b 23 and note; in Bk. XIII, Ch. 6–9, below, this conception of numbers with "incomparable" units is criticized in detail.

the first of beings is not movable either in its own right or incidentally, but sets in motion the primary motion, that is one and everlasting. But since what is in motion must be moved by something, and the first mover must be itself motionless, and an everlasting motion must be caused by an everlasting source of motion and one motion by one mover, while we see in addition to the simple motion of the whole heaven, which we claim the first motionless independent thing causes, other everlasting motions which belong to the planets (for the body that goes in a circle does so everlastingly and without stopping, which is demonstrated in the writings on nature[22]), it is necessary that each of these motions also be caused by something that is itself motionless and an everlasting independent thing. For the nature of the stars is for each to be an everlasting independent thing, while the mover is everlasting and takes precedence over the thing moved, and what takes precedence over an independent thing must be an independent thing. Accordingly, it is clear that there must be that many independent things and that they must have a nature such that they are everlasting and motionless in virtue of themselves, and, for the reason stated above, without magnitude.

1073a 30

1073b

That, then, there are independent things, and of these a first one and a second, following the same order as the motions of the stars, is evident; but the number of motions is already something one must examine from that kind of mathematical knowledge that is the nearest kin to philosophy, namely from astronomy. For this kind makes its study about perceptible, everlasting thinghood, while the others, such as those concerned with numbers and with geometry, are not about thinghood at all. Now the fact that the motions are greater in number than the things moved is clear to those who have touched on the subject even moderately (since each of the wandering stars is carried along more than one motion); but as for how many these happen to be, we now state what some of the mathematicians say, for the sake of a conception of it, in order that some definite number be grasped in our thinking, and as for what remains, it is necessary to inquire into

1073b 10

---

22 See *Physics* Bk. VIII, Ch. 8–9. Each planet has a daily westward motion that could derive from the motion of the sphere of the fixed stars, but also has a slower eastward motion belonging to it alone, which must have a source. The extra motions discussed below arise out of further anomalies, and out of the need to prevent the special motions of the planets from being passed downward through the system.

some things ourselves, while listening to what other inquirers say about others. If something should seem to those who busy themselves with these matters to be contrary to what has just now been said, it is necessary to welcome both accounts, but trust the more precise one.

Eudoxus, then, set it down that the motion of both the sun and the moon is in three spheres, of which the first is that of the fixed stars, the second rotates along a path through the midst of the zodiac,[23] and the third along a path inclined along the width of the zodiac (but with that along which the moon is carried inclined to a greater width than that along which the sun is carried); but he set it down that the motion of each of the wandering stars is in four spheres, of which the first and second are the same as the former ones (for the sphere of the fixed stars is that which moves them all, and the sphere assigned to the place under this and having its motion along a path through the midst of the zodiac is common to them all), while the poles of the third sphere for all of them are in the path through the midst of the zodiac, and the motion of the fourth is along a path inclined to the equator of the third. And he set it down that the poles of the third sphere are peculiar to the different planets, but those for Venus and Mercury are the same. Callippus set down the same arrangement of spheres as did Eudoxus, and gave the same number as he did for Jupiter and Saturn, but for the sun and the moon he thought there were two spheres still to be added if one were going to account for the appearances, and one for each of the remaining planets.

But it is necessary, in order to account for the appearances, if all the spheres are going to be fit together, that there be for each of the planets (less one) other spheres, turning backwards and continuously restoring to the same position the first sphere of the star situated next below; for only in that way is it possible for them all to produce the motion of the planets. Since, then, the spheres in which they themselves are carried are eight [for Jupiter and Saturn] and twenty-five [for the sun and moon, Mars, Venus, and Mercury], and of these one, that in which the star situated lowest is carried, does not need to be counter-turned, the spheres to counter-turn those of the first two planets will be six, while the ones to counter-turn those of the four lower ones will be

1073b 20

1073b 30

1074a

1074a 10

---

[23] Eudoxus, an older contemporary of Aristotle, seems to have been the first thinker to postulate an ecliptic, and thus reduce the sun's apparent spiral to a combination of two circles. The third motion accounts for north-south deviations from the ecliptic.

sixteen; so the number of all the carrying spheres plus those that turn backward against them will be fifty-five. But if one were not to add to the moon and the sun the motions which we mentioned, all the spheres will be forty-seven.[24] So let the number of the spheres be so many, so that it is reasonable to assume that the number of independent things which are motionless sources is also that many (for let the number that is necessary be left for more relentless people to say).

But if it is impossible for there to be any motion that is not directed into the motion of a star, and if in addition every nature and every independent thing that is unaffected and has, by virtue of itself, attained its best condition, must be regarded as an end, there could be no other nature besides these, but this is necessarily the number of independent things. For if there were others, they would be movers as final causes of motion; but it is impossible for there to be other motions besides those mentioned. And this is reasonable to assume from the things that are moved. For if what carries something is naturally for the sake of what is carried, and every motion belongs to something that is carried, no motion could be for the sake of itself or for the sake of another motion, but they are for the sake of the stars. For if there were to be a motion for the sake of a motion, the latter too would have to be for the sake of another one; so since this cannot go to infinity, there will be as an end for every motion one of the divine bodies carried through the heaven.[25] And it is clear that there is one heaven. For if there were a plurality of heavens, as there is of human beings, there would be one *kind* of source for each one, but many of them in number. But those things that are many in number contain material (for one and the same articulation belongs to many things, as does the articulation of a human being, but Socrates is one). But what it primarily is for something to

1074a 20

1074a 30

---

[24] This appears to be a mistake by Aristotle. He has subtracted the four spheres added by Callipus, and four more that would have been their counter-turners, forgetting that the lowest (the moon) had no counter-turners. His joke in the next sentence suggests that some student caught his error, but he left it in.

[25] The argument says that there are no lost motions, no invisible effects, and hence no undetectable sources. But if it is correct that the main argument of Bks. VII–IX converges into Bk. XII (with a stepping stone in Bk. X, Ch. 1), these same sources that move the stars are the simple, active beings responsible for the unity of the forms of living things. In that way they work directly upon earthly material, forming it into wholes. The simplicity and constancy of the stars imitates them by final causality, making their presence available to eyesight, as well as to philosophic reasoning.

be does not contain material, for it is a being-at-work-staying-itself. Therefore the first[26] motionless being that causes motion is one both in articulation and in number; and therefore what is moved always and continuously is also one. Therefore there is only one heaven.

There has been handed down from people of ancient and earliest   1074b
times a heritage, in the form of myth, to those of later times, that these original beings are gods, and that the divine embraces the whole of nature. The rest of it was presently introduced in mythical guise for the persuasion of the masses and into the laws for use and benefit; for the myths say the gods are of human form or like some of the other animals, and other things that follow along with and approximate these that have been mentioned. If one were to take only the first of these things, separating it out, that they thought the primary independent things were gods, one would regard this as having been said by divine   1074b 10
inspiration, and, since it is likely that every kind of art and philosophy has been discovered to the limit of its potential many times, and passed away in turn,[27] one would consider these opinions of those people to have been saved like holy relics up to now. So the opinion of our forefathers that comes from the first ages is clear to us but only to this extent.

Chapter 9 Now concerning the intellect there are certain impasses, for it seems to be the most divine of the things that are manifest to us, but the way it is if it is to be of that sort contains some things that are hard to digest. For if it thinks nothing, what would be solemn about that? Rather, it would be just like someone sleeping. But if it does think, but something else has power over it, then, since it is not   1074b 20
thinking but potency that is the thinghood of it, it could not be the best independent thing, for it is on account of its act of thinking that its place of honor belongs to it. And still, whether the thinghood of it is a power of thinking or an activity of thinking, what does it think?

---

[26] The mover of the first and all-encompassing sphere must be meant. It must be single and simple, though the plurality of heavenly motions reflects that the whole of thought thinking itself contains multiplicity. It suggests a divine community, with a dominant member.

[27] Since past time is unlimited, human arts and knowledge must have emerged fully; since they are defective, they must have been lost again at some time. Progress is only a sign of previous catastrophe.

For this is either itself or something else, and if it is something else, either always the same one or different ones. And then does it make any difference, or none, whether its thinking is of what is beautiful or of some random thing? Isn't it even absurd for its thinking to be about some things? Surely it is obvious that it thinks the most divine and honorable things, and does not change, since its change would be for the worse, and such a thing would already be a motion. First, then, if it is not an activity of thinking but a potency, it is reasonable to suppose that the continuation of its thinking would be wearisome; and next,

1074b 30   it is clear that something else would then be more honorable than the intellect, namely what it thinks. For thinking and the activity of thinking would belong even to something that thinks the worst thing, and if this is to be avoided (for it can even be more advantageous not to see some things than to see them), then the activity of thinking would not be the best thing. Therefore what it thinks is itself, if it is the most excellent thing, and its thinking is a thinking of thinking.[28]

But knowledge and perception and opinion and step-by-step thinking seem always to be about something else, and about themselves only as something secondary. What's more, if the thinking and the being thought are different, then in virtue of which of them does what is good belong to it? For to be an act of thinking and to be something

1075a   thought are not the same. Or is it rather that in some cases the knowledge is the thing it is concerned with, so that in the case of the kinds of knowing that make something, the thinghood without material and what it is for something to be, or in the case of the contemplative kinds of knowing, the articulation, is both the thing the knowledge is concerned with and the activity of thinking it? So since what is thought and what is thinking are not different with as many things as have no material, they will be the same, and the act of thinking will be one with what is thought.

But there is still an impasse left as to whether what is thought is composite, for then the thinking would be changing among the parts

---

[28] This is not the structure of self-consciousness but just the opposite. The activity of the divine intellect is nothing over and above its content, and none of its content is unassimilated into its activity. It is not a doubled thinking that turns around to look at itself, but a fusion of activity with that upon which it works. Thus it could be described equally well from the side of its content, as form informing itself, or forms maintaining themselves in being by their own act.

of the whole. Or is it the case that everything that has no material is indivisible? So the condition the human intellect, or that of any composite being, is in at some period of time (for it does not have hold of what is good at this or that time, but in some whole stretch of time it has hold of what is best, since that is something other than itself), is the condition the thinking that thinks itself is in over the whole of time.[29]

1075a 10

Chapter 10 One must also consider in which of two ways the nature of the whole contains what is good and what is best, whether as something separate, itself by itself, or as the order of the whole of things. Or is it present in both ways, just as in an army? For its good condition resides in its ordering but also is its general, and is more the latter; for he does not depend on the order but it on him. And all things are in some way ordered together, though not all similarly, the things that swim and fly and grow in the ground; yet they are not such that nothing that pertains to one kind is related to another, but there is some relation. For they are all organized toward one thing, but in the same way as in a household, in which the free members of it are least of all allowed to do any random thing, but all or most of what they do is prescribed, while for the slaves and livestock little that they do is for the common good and much is just at random, since the nature of each of them is that kind of source. I mean, for example, that it necessarily comes to everything at least to be decomposed, and there are other things of this sort that all things take part in together that contribute to the whole.

1075a 20

But how many impossible or absurd consequences follow for those who speak of these things in other ways, and what sort of things those who speak of them most gracefully say, and what sort of opinions have the fewest impasses, must not be ignored. For everyone makes all things come from contraries, but neither the "all things" nor the "from contraries" is right, and even with those things in which contraries are present, they do not say how they come from those contraries, for

1075a 30

---

[29] There is an analogy in the two ways we might look at a landscape or a painting. At first our attention goes from part to part, and we see the whole only in memory or imagination, and over a stretch of time. But we can also let our eyes drink in a scene which is all there for them all along. Such an experience is out of time, and takes in as simple what is in other respects composite.

of being badly governed. "A divided sovereignty is not good; let there be one lord."[35]

---

[35] With these words at *Iliad* II, 204, Odysseus turns the Achaean multitude back from a chaotic rout and into an orderly army. Similarly, the divine intellect described by Aristotle does not create things or the world, but confers upon them their worldhood and thinghood.

Book XIII (Book M)
The Being of Intelligible Things[1]

Chapter 1 About the thinghood of perceptible things, what it is has been said, with respect to material in the pursuit that deals with natural things, and later with respect to the thinghood that has being as being-at-work.[2] But since the investigation is whether there is or is not anything apart from perceptible independent things that is motionless and everlasting, and if there is, what it is, one must first consider the things said by others, in order that, if they say anything that is not right, we might not be subject to the same mistakes, and if some teaching is common to us and them, we might not be annoyed that this does not peculiarly come from us; for one ought to be content if one says some things better and others not worse.

And there are two opinions about these things, for some people say that mathematical things, such as numbers and lines and things of a like kind with these, are independent things, and again that the forms are independent things. But since some make these two kinds, the forms and the mathematical numbers, while others make one nature for both, and yet others say that the mathematical things alone are independent things, one must investigate first about the mathematical things, not adding any other nature to them, such as whether they turn out to be forms or not, or whether they are sources and the thinghood of beings or not, but as concerns mathematical things alone, whether they *are* or are not, and if they are, in what way they are. Next after this, one must investigate separately about the forms themselves simply and enough to satisfy custom, for these have been blathered about in many ways and in popular writings, but also the bulk of the account must be additional to that whole investigation, when we come to examine whether the thinghood and the sources of beings are numbers and forms, for after the forms are discussed this third investigation remains.

1076a 20

1076a 30

Now it is necessary, if mathematical things *are*, that they be either in the perceptible things, as some people say, or separate from the

---

[1] This title for Book XIII supplied by the translator.

[2] The pursuit that deals with natural things is the *Physics;* the identification of form with being-at-work occurs primarily in Book VIII above.

perceptible things (and some people also speak of them that way); or if they are not present in either way, then they either do not have being or they have it in some other manner. So for us the dispute will not be about whether they have being, but about the *manner* of their being.

Chapter 2 That it is impossible for the mathematical things to be *in* the perceptible things, along with the fact that this account is a source of deception, has been said in the discussion of impasses,[3] because it is impossible for two solids to be coincident, and also because according to the same argument the rest of the capacities and natures would be in perceptible things and none would be separate. These things have been said before, but in addition to these it is clear that it would be impossible for any body at all to be divided, since it would be divided at a surface, and this at a line, and this at a point, so that, if the point cannot be divided, then neither can the line, and if that cannot, then neither can the rest. What difference does it make, then, whether these perceptible bodies are such natures, or are not but have such natures in them? For the same thing will result, since they will be divided when the perceptible things have been divided, or else not even the perceptible things will be divided.[4]

But neither is it possible for such natures to be separate. For if there were separate solids besides the perceptible solids, different from and taking precedence over the perceptible ones, it is clear that, besides the perceptible surfaces, there must be other separate surfaces and points and lines (since this comes from the same argument); but if there are these, then again besides the surfaces of the mathematical solid there would also be other separate lines and points. (For uncompounded things take precedence over composite ones, and if bodies that are not perceptible take precedence over the perceptible ones, then by

1076b

1076b 10

1076b 20

---

[3] See 998a 7–19.

[4] In the *Physics*, in his final refutation of Zeno's paradoxes (263a 4–263b 9), Aristotle argues that a dividing point or line is present in a continuous thing only potentially, until the division is made or at least singled out for attention. If the mathematical things are already in the divisible body, there must be an actual rather than a potential infinity of them. In a similar way, in the argument referred to just above, no "solidity" prevents a mathematical solid from coinciding with a perceptible body, but neither is the former a second thing "in" the latter, but a potency of the latter, by which it can be attended to as though it lacked its perceptible attributes. See 1061a 29–1061b4, and Ch. 3 of this book.

the same argument, also surfaces that are just by themselves would take precedence over the surfaces in motionless solids, so that these surfaces and lines would be different from the ones that are together with the separate solids; so there is one sort that are together with the mathematical solids, and another sort that take precedence over the mathematical solids.) Then again, there would be lines belonging to these surfaces, taking precedence over which there would have to be other lines and points, by the same argument, and taking precedence over the points belonging to these lines that take precedence, there would be other points, but no longer would there be anything taking precedence over these. So the piling up becomes absurd. (For there turn out to be one batch of solids besides the perceptible ones, three batches of surfaces besides the perceptible ones—those besides the perceptible surfaces, those in the mathematical solids, and those besides the surfaces in these mathematical solids—four batches of lines and five of points; so then which of these kinds is mathematical knowledge about? For it is surely not about the surfaces and lines and points in the motionless solid, since knowledge is always about what takes precedence.) And the same argument also applies to numbers, for there would be a different batch of units besides each batch of points, and also besides each batch of beings, the perceptible ones and then the intelligible ones, so that there would be kinds of mathematical numbers.

1076b 30

Further, how do those things we went over among the impasses[5] admit of being resolved? For the things about which astronomy is concerned would similarly be apart from the perceptible things, as would the things about which geometry is concerned. But how is it possible for there to be a heaven and its parts aside from the perceptible ones, or anything else whatever that has motion? And likewise with the things studied by optics and harmonics, for there would be sound and sight apart from the ones that are perceptible and particular, and so it is clear that it would be the same with the other senses and the other perceptible things, for why these rather than those? And if there are these things, there would also be animals, if indeed there are senses. Also, some things are treated by mathematicians in a universal way,[6]

1077a

1077a 10

---

[5] See 997b 14–32.

[6] In Book V of Euclid's *Elements*, ratios of magnitudes are treated in a general way;

apart from these independent things, and thus there would also be this other kind of in-between thinghood separate from both the forms and the things in-between the forms and the perceptible things, which is neither number nor points nor magnitude nor time. But if this is impossible, it is clear that it is impossible also for those things to be separate from the perceptible ones. In general, things turn out to be contrary both to the truth and to what it is customary to assume if one posits that mathematical things *are* in this way, as certain separate natures. For while it is necessary, on the supposition of their being in that way, that they take precedence over the perceptible things, in truth they are derivative; for while, in the order of coming into being,

1077a 20   the incomplete magnitude is prior, in thinghood it is derivative, just as what is without soul is derivative from what has soul.[7]

Also, how and when would mathematical magnitudes be one? For the things around us are one by virtue of a soul, or a part of soul, or something else, reasonably (and if not they are many, and come apart), but with those things that are divisible and are quantities, what is responsible for their being one and holding together? Also, the coming into being of a mathematical magnitude makes this clear. For first it comes to be along its length, then along its width, and last into its depth, and it has its completion. So if what is later in coming into being is prior in thinghood, body would be prior to surface and length; and also a body is more complete and whole in the sense that it is what comes

1077a 30   to be ensouled, but how could a line or surface be ensouled? To think it possible would go beyond the power of our senses. Also, a body is a certain sort of independent thing (for it already has completeness in a certain way), but how could lines be independent things? Neither as a certain form or look, if, for example, that is why the soul is an independent thing, nor as material, in the way the body is, for nothing

---

this passage and the beginning of Chapter 3 below refer to theorems that would encompass ratios of continuous magnitudes and of numbers. But no new object arises to correspond to this generality of treatment; the things in the ratios must be either numbers or definite magnitudes, even though the theorem's truth is indifferent to their particular natures. The symbolic procedure of algebra is an entirely different approach to universality in mathematics from what Aristotle is describing. See Jacob Klein, the book cited in the footnote to 987b 23.

[7] Mathematical magnitudes are incomplete in that they are stripped down from perceptible ones, just as the corpse of an animal is incomplete and lacking what gives it being.

is evident that is capable of being composed of lines or surfaces or points, but if they were some sort of material thinghood, there would have been evidence that things could have this happen to them.

And let them be prior in articulation; still, not all things that are prior in articulation are prior in thinghood. For those things are prior in thinghood which, when separated from other things, surpass them in being, while things are prior in articulation to those things of which the articulations come from their articulations, and these do not belong to things together. For if attributes, such as being in motion in some way, or white, do not have being apart from independent things, white is prior to the white human being in articulation but not in thinghood, since it does not admit of being separated but is always together with the composite whole (and by the composite whole I mean the white human being). So it is clear that the thing that results from taking something away[8] is not prior, nor is the thing that results from adding something derivative, since it is by an addition to the white that the white human being is articulated.

It has been said sufficiently, then, that mathematical things are not independent things more than bodies are, nor are they prior in being to perceptible things, but only in articulation, nor are they capable of being somewhere as separate; but since they are not capable of being *in* the perceptible things either, it is clear that they either have no being at all, or that they have being in a certain manner and for this reason do not have being simply, for we speak of being in a number of ways.

Chapter 3 Now just as the things that are universal within mathematics are not about things that are separate from magnitudes and numbers, but are about these, but not insofar as they are of such a sort as to have magnitude or to be discrete, it is clear that it is also possible for there to be both articulations and demonstrations about perceptible magnitudes, not insofar as they are perceptible but insofar as they are of certain sorts. For just as there are also many discourses about things insofar as they are in motion alone, separate from what each such thing is and from the attributes that belong to them, and there is no need on

1077b

1077b 10

1077b 20

---

[8] This is the word sometimes translated as "abstraction." Aristotle uses it only in the simple sense it has here, never as the source of the forms, and always in connection with mathematical things.

this account for there to be any moving thing separate from the perceptible ones or for there to be any distinct nature within these, so too in the case of moving things there will be discourses and knowledge, not insofar as they are in motion but insofar as they are bodies alone, and again insofar as they are surfaces alone and insofar as they are lengths

1077b 30   alone, and insofar as they are divisible or insofar as they are indivisible but having position, and insofar as they are indivisible alone; so since it is true to say simply that there are not only separate things but also things that are not separate (as there are moving things), it is also true to say simply that there are mathematical things and that they are of such a sort as people say. And just as it is also true to say of the other kinds of knowledge that they are simply about this or that and not about what incidentally belongs to it (for instance that one of them is about the white, if a healthy thing is white, when it is about health), but

1078a   about that with which each is concerned, with health if it is health, but with the human if it studies it insofar as it is human, so too is it the case with geometry. If it is about things which incidentally are perceptible, but is not concerned with them insofar as they are perceptible, mathematical knowledge will not be about perceptible things; however, it will not be about other separate things besides these either.

Many things go along with things in virtue of themselves insofar as each belongs among things of a certain sort, since also, insofar as an animal is female or insofar as it is male, there are attributes peculiar to it (even though there is nothing female or male separate from animals), and so too insofar as things are lengths alone or insofar as they are sur

1078a 10   faces. And to the extent that something concerns things that are prior in articulation and are simpler, to that extent it will have more precision (for this is what is simple), and so what is without magnitude is simpler than what has magnitude, and what is simplest is without motion but if it has motion it is most so if it has the primary motion, since that is the simplest motion, and simplest of this if it is uniform. The same argument also applies to harmonics and optics, for neither of them studies anything insofar as it is a sight or a sound, but insofar as things are lines and numbers (but these are attributes proper to those things), and likewise with mechanics; and so if someone examines anything concerning these attributes, insofar as they are such, positing that they are separate, he will not on this account cause anything to be false, any more

1078a 20   than when one draws a line on the ground that is not a foot long and says it is a foot long, for the false assumption is not in the proposition.

The best way to study each thing would be in this manner, if one were to posit as separate what is not separate, the very thing that the arithmetician and the geometer do. For a human being, as a human being, is one and indivisible, and the person who has posited this as one indivisible thing then considers whether anything goes along with the human being insofar as it is indivisible. But the geometer takes it neither as a human being nor as indivisible, but as a solid. For it is clear that what would have belonged to it anyway, even if it were not indivisible, can also belong to it along with this, and so for this reason the geometers speak rightly, and talk about things that *are* and are beings, since being is of two sorts, the one sort fully at-work and the other in the manner of material.[9]

1078a 30

And since the good and the beautiful are different (for the former is always involved in action but the beautiful is also present in motionless things), those who claim that the mathematical kinds of knowledge say nothing about what is beautiful or good are wrong. For they speak of it and reveal it most of all, for if they do not name it but their deeds and discourses make it evident, it is not the case that they do not speak about it. The greatest forms of the beautiful are order and symmetry and determinateness, which the mathematical kinds of knowledge most of all display. And since these make their appearance as causes of many things (I mean such things as order and determinateness), it is clear that these kinds of knowledge would also speak about that which has responsibility in the manner of the beautiful as a cause in some manner. But we will speak more explicitly about these things in other places.[10]

1078b

Chapter 4 So about the mathematical things, that they are beings and in what manner they are beings, and in what way they are prior and in what way not prior to other things, let so much have been said. But concerning the forms, one ought first to examine the opinion about form itself, not connecting it with the nature of numbers, but in the way the first people who spoke about the forms at first conceived them to

1078b 10

---

[9] To be "in the manner of material" is to be a potency. (See 1050a 18.) One might say that potencies are "in" things, but Aristotle takes that preposition as implying active presence.

[10] This refers to Bk. XIV, Ch. 4, but also applies to Bk. XII, Ch. 7 and 10.

be.[11] The opinion about the forms came to those who spoke about them as a result of being persuaded by the Heracleitean writings that it is true that all perceptible things are always in flux, so that, if knowledge and thought are to be about anything, there must be, besides the perceptible things, some other enduring natures, since there can be no knowledge of things in flux. And then Socrates made it his business to be concerned with the moral virtues, and on account of them he first sought to define things in a universal way. (For among those who studied nature, only

1078b 20   to a small extent did Democritus attain to this and define in some way the hot and the cold; and before that the Pythagoreans did so about some few things, the articulations of which they attached to numbers, as with what due measure is, or the right way, or marriage. But it was reasonable that Socrates sought after what something is, for he was seeking to reason deductively, and the starting point of deductive reasoning is what something is; for dialectic was then not yet well-developed enough to be able, apart from what something is, to examine it from contrary assumptions. For there are two things one might justly credit Socrates with, arguments by example and universal definition,

1078b 30   for both of these are approaches to the starting point of knowledge.) But Socrates did not make the universals or the definitions separate, while those who came next did, and called beings of this sort forms; so for them it followed by pretty much the same argument that there are forms of all things that are spoken of in a universal way,[12] and it is just about as if someone who wanted to count a smaller number of things thought he couldn't do it, but could count them if he made

1079a   them more. For the forms are, in a manner of speaking, more numerous than the separate kinds of perceptible things, in quest of the causes of which they had moved on from those things to the forms; for there is a form for each thing for which there is a name, besides the independent things, and for the other ways of being there is also a one applied to the many, both for the things around us and for everlasting things.

---

[11] These first opinions about forms are those evident in Plato's dialogues, as a comparison of the following passage with Bk. I, Ch. 6, confirms. The later opinions, connecting forms with numbers, also involve Plato and various students and successors of his in the Academy.

[12] From this point to the end of Chapter 5, most of the text repeats part of Bk. I, Ch. 9, from 990b 1 to 991b 9.

Also, by the ways in which they show that there are forms, by none of them is this evident, for from some arguments no conclusion necessarily follows, and from some there turn out to be forms of those things which they believe to have none. By the arguments from the kinds of knowledge, there would be forms of all things of which there is knowledge, by the argument for one thing applied to many there would be forms even of negations, and by the argument from thinking of something that has been destroyed, there would be forms of things that have passed away, since there is some imagining of these. Also, some of the most precise of the arguments produce forms of relations, which they claim do not make up a class of things in their own right, while others imply the third man.[13] And in general, the arguments about the forms abolish things which those who speak of the forms want there to be more than they want there to be forms. For it turns out that not the dyad but number is primary, and of number the relative sort is primary, and the relative is more primary than what *is* in its own right, and all those things turn out to be the case that certain people, who followed out the opinions about the forms to their logical conclusions, showed to be opposed to the original sources of things.[14]

1079a 10

Also, by the assumption by which they say there are forms, there would be forms not only of independent things, but also of many other things (since the thing thought is one not only as concerns independent things but also as applied to what are not independent things, and knowledge is not only concerned with independent things, and there turn out to be vast numbers of other such things).[15] But both necessarily and as a result of the opinions about them, if the forms are shared in, there must be forms only of independent things. For a thing does not share in a form incidentally, but must share in each by virtue of that in respect to which it is not attributed to an underlying thing. (I mean, for example, if something partakes of double itself, then it also partakes of the everlasting, but incidentally, since it is incidental to the double to be everlasting.) Therefore the forms will be the thinghood of things,

1079a 20

1079a 30

---

[13] See footnote to 990b 17.

[14] See footnote to 990b 24.

[15] Any possible attribute is one in thought as against its many instances, so there would not just be a form of apple (itself not strictly an independent thing, since it is maintained in being only as part of an apple tree), but also of red, of rounded, of juicy, etc.

and the same things will signify thinghood there as here; otherwise what would the something be that is said to be apart from the things around us and that is the one over the many? And if the forms and the things that participate in them have the same form, there would be something common. (For why is *two* one and the same thing as applied to both destructible pairs and the pairs that are many but everlasting, any more than as applied both to itself and to something?) But if the form is not the same, there would be ambiguity, and it would be just as if someone were to call both Callias and a block of wood "human being" while observing nothing common to them.

1079b

But if we posit that the common articulations are well suited to the forms in other respects, for example, as applied to the circle itself, that it is a plane figure and the rest of the parts of the definition, but that "that which *is*..." be added, one must examine whether this is not completely empty. For to what would the addition be made, to "center" or to "plane" or to all the parts? For everything in its thinghood is a form, as are both animal and two footed. It is also clear that the added part must itself be something, as a plane is, some nature which would be present in all the rest as a genus in its species.

1079b 10

Chapter 5 But most of all, one might be at a loss about what in the world the forms contribute to perceptible things, either to the everlasting ones or to the ones that come into being and perish. For they are not responsible for any motion or change that belongs to them. But they don't help in any way toward the knowledge of the other things either (for they are not the thinghood of them, since in that case they would be in them), nor toward their being, inasmuch as they are not present in the things that partake of them. For they might perhaps seem to be causes in the manner of the white that is mingled in a white thing, but this argument, which Anaxagoras made first, and Eudoxus made later as a paradox, and some other people made, is truly a pushover (for it is easy to collect many impossibilities related to this sort of opinion). But surely it is not true either that the other things are made out of the forms in any of the usual ways that is meant. And to say they are patterns and the other things participate in them is to speak without content and in poetic metaphors. For what is the thing that is at work, looking off toward the forms? And it is possible for anything whatever to be or become like something without being an image of it, so that whether Socrates is or is not, one might become like Socrates, and it is

1079b 20

1079b 30

obvious that it would be the same even if Socrates were everlasting. And there would be more than one pattern for the same thing, and so too with the forms; for example, of a human being, there would be animal and two footed, as well as human being itself. What's more, the forms will be patterns not only of the perceptible things but also of themselves, such as the form *genus,* since it is a pattern of the forms in that each is a genus, and so the same thing would be both a pattern and an image.

Further, one would think it was impossible for the thinghood and that of which it is the thinghood to be separate; so how could the forms be the thinghood of things if they are separate? In the *Phaedo* it is put this way: that the forms are responsible for both being and becoming. And yet even if there are forms, still the things that partake of them do not come into being if there is not something that causes motion, and many other things do come into being, such as a house or a ring, of which they say there are no forms. So it is clear that the other things too, of which they say there are forms, admit of being and becoming on account of the sort of causes that are responsible for the things just mentioned, and not on account of the forms. But concerning the forms, both in this way and by arguments from a more logical standpoint and of a more precise character, it is possible to collect many things like the ones that have been examined.

Chapter 6 Since boundaries have been drawn around these things, it would be good to consider next in turn[16] the things that turn out to be the case concerning numbers for those who say that they are separate independent things and the first causes of beings. And necessarily, if number is a certain nature, and the thinghood of it is not anything else but just this, as some people say, then either there is something that is first in it and something else next, each of them being different in kind, and this either belongs to the units directly, so that any unit is incomparable with any other unit, or the units are all in direct series and any is comparable with any other, the sort of thing people say a

1080a

1080a 10

1080a 20

---

[16] Mathematical things have been considered in Chapters 2–3, and the forms have been considered in a nontechnical way in Chapters 4–5. Chapters 6–9 deal with a special technical conception that fuses the forms with numbers. This was a topic of discussion in Plato's Academy, aimed at working out an understanding of the forms as having a numerical structure. See the footnote to 987b 23.

mathematical number is (since in mathematical number one unit does not differ in any way from another unit), or some are comparable and others not.[17] (For example, if two is the first number after one, then three, and so on with the other numbers, and the units within each number are comparable, that is, those in the first two comparable among themselves and those in the first three among themselves, and so on with the other numbers, while those in the two itself are incomparable in relation to those in the three itself, and similarly with

1080a 30   the other numbers in series, then for this reason, while mathematical number is counted with two after one, with another one added to the one before it, and three with another one added to these two, and the rest in the same way, this sort of number is counted with a different two after one, without the first one in it, and a three without the two, and similarly with the other numbers.) Or there is one sort of number of the kind first described, another sort that is as the mathematicians say, and a third sort that was mentioned last. It is also necessary that

1080b   these numbers be either separate from things, or not separate but in the perceptible things (not in the sense that we first considered, but in the sense that the perceptible things are made of numbers as constituents of them[18]), either some numbers in things and others not or all of them in things.

These, then, are necessarily the only ways in which it is possible for numbers to have being, and of those who say the one is the source and thinghood and element of all things, and that number comes from this along with something else, just about each one of them has described number in some one of these ways, except the one that makes all the

1080b 10   units be incomparable. And this turns out reasonably, since it is not possible for there to be any other way besides those mentioned. Some people say that numbers are of both sorts, one with the forms holding places before and after, the other, the mathematical sort, besides the forms and the perceptible things, with both sorts separate from the perceptible things; others say that it is mathematical number alone that is primary among beings, separate from the perceptible ones. The

---

[17] The Greek word for number meant primarily any multitude of things, and only secondarily a multitude of indifferent units. The first and third alternatives here refer to ways that the forms themselves might be assembled in groups and an order.

[18] The sense first considered was at the beginning of Ch. 2 above; the new sense is the Pythagorean one, discussed below.

Pythagoreans say it is of one sort, the mathematical, except they say it is
not separate but that perceptible independent things are composed of
this, for they make the whole cosmos out of numbers, but not numbers
of arithmetic units, but they assume that units have magnitude; how
the first *one* was composed as having magnitude, they seem to have no
way of saying. And there is someone else who says the primary sort of
number is the one composed of forms, and some people say that the
mathematical sort is this same thing.

    And it is similar also concerning lengths and surfaces and solids.
For some say the mathematical sort are different from the sort that are
together with the forms; but of those speaking otherwise, some say
they are mathematical things and treat them in a mathematical way,
and these are the people who do not make the forms numbers or say
that these things are forms, while others say they are mathematical
things but do not treat them in a mathematical way, for they say that
not every magnitude can be cut up into magnitudes, and not any and
every pair of units can be two.[19] All those who say that the one is an
element and source of beings posit that there are numbers composed
of arithmetic units, except for the Pythagoreans, and they say the units
have magnitude, as was said before. In how many ways, then, numbers
admit of being spoken about, and that the ways described are all of
them, are clear from these things; and there are impossibilities with all
of them, though perhaps more with some than with others.

    Chapter 7 First, one must examine whether units are comparable
or incomparable, and if they are incomparable, in which of the two
ways we distinguished. For it is possible for any unit to be incomparable with any other unit whatever, or it is possible for those within two
itself to be incomparable with those within three itself, and so on with
those within each primary number in relation to each other primary
number. And then if all units are comparable and undifferentiated,
number comes to be mathematical and of one sort only, and it is not
possible for the forms to be numbers. (For what sort of number would
human being itself be, or animal itself, or any at all of the other forms?
For there is one form of each, that is, one of human being itself, and

1080b 20

1080b 30

1081a

1081a 10

---

[19] Scholars identify Speusippus among the "some" and Xenocrates among the
"others." The "someone else" in the last paragraph is unknown.

another one of animal itself, but the units that are alike and undifferentiated are infinite, so that human being itself could no more be this or that three than any other.) But if the forms are not numbers, it is not possible for there to be forms at all. (For from what sources would the forms be? For number is from the one and the indeterminate dyad,[20] and the sources and elements are spoken of as being *of* number, and so the forms would not admit of being ordered as prior to or derivative from the numbers.)

But if the units are incomparable, and in such a way that any at all is incomparable with any other at all, then it is not possible for this sort of number to be mathematical. (For mathematical number is made of undifferentiated units, and the things demonstrated about it are fitted to it as being of that sort.) But neither is it possible for it to be the sort of number the forms are; for two would not be the first number to come from the one and the indeterminate dyad, nor would the series of numbers spoken of as two, three, four be next, for the units in the primary two are generated together (whether in the way the first one who spoke of this said, from unequals, since they were generated when they became equal, or in some other way), since if one unit were prior to another, it would also be prior to the two made out of these, for whenever anything has one part prior and another derivative, that which is made of these will be prior to the latter and derivative from the former.[21] Also, since the one itself is first, and then of the rest there is a one that is first and a second one after that, and then a third that is second after the second one and third after the first, therefore the units would be prior to the numbers from which they are named, that is, the third unit would be present in the number two prior to there being a number three, and the fourth and fifth would be in the number three prior to these numbers.

Now none of them has said that the units are incomparable in this manner, but it is reasonable according to their starting assumptions

1081a 20

1081a 30

---

[20] See the work by Jacob Klein cited in the footnote to 987b 23. The *one* understood as a form and a source is not a number but that which confers limit; the indeterminate dyad (a technical term equivalent to "the great and the small") is undifferentiated muchness which, when limited by the one, becomes the determinate dyad, the eidetic number two, interpreted as the form of being.

[21] If the units within a single form are incomparable, they are ranked in order before and after the number itself, and destroy the ordering of the forms as numbers.

that they be so even in this way; according to the truth, however, it is impossible. For it is reasonable that the units be prior and derivative if there is also a first unit and a first *one,* and similarly that the twos be prior and derivative if there is a first two, since after a first it is reasonable and necessary that there be a second, and if a second, a third, and so on with the rest in series (but to say both at the same time, that there is a first and a second unit after the one, and also that there is a first two, is impossible). But they make a unit and a first one, but no longer a second and a third, and a first two, but no longer a second and a third.

And it is also clear that it is not possible, if all the units are incomparable, for there to be a two itself and a three and the other numbers of that sort, for not only when the units are undifferentiated but also when they differ each from each, it is necessary for number to be counted by addition, as two by adding to one another one and three by adding another one to two and four in a similar way; but if this is so, it is impossible for there to be a coming into being of the numbers in the way they generate them from the dyad and the one. For two becomes part of three, and this becomes part of four, and it follows in the same way with the succeeding numbers. But they have four coming from the first two and the indeterminate dyad, two dyads that are other than two itself. If that is not so, two itself will be part of four, and one other two will be added, and two will come from one itself and another one; but if this latter is so, it is not possible for the second element to be the indeterminate dyad, since that second element brings about one unit and not a determinate two. Also, how will there be other threes and twos besides three itself and two itself? In what manner are they put together out of prior and derivative units? For all these things are absurd and like fiction, and it is impossible that there be a first two and then three itself, but this is necessary so long as the one and the indeterminate dyad are the elements. But if the consequences are impossible, then these starting points are impossible.

So if any whatever of the units differs from any other whatever, these things and others of this sort necessarily follow, and if the units in different numbers differ, while those within the same number are undifferentiated from one another, even so consequences follow that are no less inconvenient. For, say, in ten itself there are ten units, but ten is composed both of these and of two fives. But since ten itself is not any chance number nor is it composed of any chance fives, nor of

1081b

1081b 10

1081b 20

1081b 30

1082a

number, these do differ in amount. But if the units too were to differ in amount, then a number could differ from a number equal to it in multitude of units. And are the first units greater or smaller, and do the later ones increase or the reverse? All these things are without reason. But surely it is not possible for them to differ in kind either, since no attribute is capable of belonging to them, for even these people say that being of a certain sort belongs to the numbers as something derivative from their amount. Also, this could not come to them from the one or from the dyad, for the former is not of any sort and the latter is a source of quantity, since this latter nature is responsible for beings' being many. So if things are otherwise in any way, one is obliged to say this especially at the beginning, and to mark out, concerning the difference in a unit especially, why it is even necessary for there to be one; if it is not, what are they talking about?

Now if the forms are numbers, it is clear that it is not possible for all the units to be comparable, and that it is not possible either for them to be incomparable with one another in either of two ways. But surely neither is the way some others speak about the numbers right. These are the people who do not believe there are forms, neither simply nor as being some sort of numbers, but believe there are mathematical things, and that numbers are the primary beings, and that the source of them is the one itself. For it is absurd for there to be a certain *one* that is the first of the ones, just as the others say, but not a two that is first of the twos or a three that is first of the threes, since they all pertain to the same argument. So if the things that pertain to number are so, and someone posits that there are only numbers of a mathematical sort, it is not possible for the one to be a source (for this sort of *one* must differ from the other units, and if that is so, there must also be some two that is first of the twos and similarly also with the other numbers in series); but if the one is a source, then it is necessary instead that what pertains to numbers is the way Plato said it is, and that there are a first two and three, and that the numbers are not comparable with one another. But again, if one posits these things in turn, it was said that many impossible things follow. But surely it has to be one way or the other—or if it is neither way, number could not admit of being something separate.

It is clear from these things also that the worst things that are said are those of the third kind, that make number as the forms and mathematical number be the same. For two mistakes come together

1083a 10

1083a 20

1083a 30

1083b

into the one opinion; for mathematical number cannot be this way, but it is necessary for the one supposing it is to spin out assumptions peculiar to it, and also to have to say all the things that follow for those who speak of number as forms. The way taken by the Pythagoreans in one respect has fewer inconveniences than those mentioned before, but in another respect has others peculiar to it. For in consequence of not making number separate, it takes away many of the impossibilities, but in consequence of making bodies be composed of numbers, and making this number be the mathematical sort, it is impossible. For to speak of indivisible magnitudes is not true, and even if they were this way as much as one wished, still the units would not have magnitude, and in what way is a magnitude capable of being composed of indivisible things? But surely mathematical number is composed of arithmetic units, but these people say that beings are numbers. At any rate they apply theorems to bodies as though they were made of those numbers. If, then, it is necessary, so long as number is some sort of being in its own right, for it to be in some one of the ways mentioned, and none of these is possible, it is clear that there does not belong to number any such nature as those who make it separate build up for it.

1083b 10

1083b 20

Also, does each unit come from the great and the small when they are made equal, or one from the small and one from the great? If it is the latter way, then each thing is not made of all the elements, nor are the units undifferentiated (for the great is present in one and the small in another, which are contrary in nature). And what are the units like in three itself? For there is one extra; but perhaps for this reason they put the one itself in the middle in an odd number. But if each of the units is from both sources made equal, how will two be some one nature which is made of the great and the small? Or how will it differ from a unit? And the unit is prior to two (since if it were destroyed, two would be destroyed), so then it must be a form of a form, since it is prior to a form, and must have come into being before it. But from what? For the indeterminate dyad makes things double.

1083b 30

Also, it is necessary that number be either infinite or finite, since they make number separate, so that it is not possible for neither of these to belong to it.[22] But it is clear that it cannot be infinite. (For

1084a

---

[22] According to Aristotle, number itself is a potency of things to be counted (see

the infinite is neither odd nor even, while the coming into being of numbers is always of either an odd or an even number—odd when in one way the one inserts itself into an even number, but in the series of doubles from one when in another way the dyad factors itself in, or one of the other even numbers when it factors itself into the odd numbers. Also if every form is a form of something and the numbers are forms, then also the infinite will be a form of something, either in perceptible things or in something else; yet neither according to the hypothesis nor according to reason is this possible, for those who put the forms into order in this way.)

1084a 10

But if number is finite, what amount does it go up to? For one ought to say not only that this is so, but why it is so. But surely if numbers go up to ten, as some say, then first, the forms will quickly run out; for example, if three is human being itself, what number is horse itself? For each thing itself is a number up to ten, so it must be one of the numbers among these (for these are the kinds of thinghood and the forms); however, they will run out (for the forms of animals will exceed them). At the same time it is clear that if in this way three is human being itself, then the other threes are too (for the units in the same numbers are alike), so human beings would be infinite, and if each three is a form, each one would be human being itself, but if not, they would still be human beings. And if a lesser number is part of a greater number, the number being the sort made of units that are comparable within the same number, then if four itself is a form of something, say of horse or of white, a human being will be part of a horse, if human being is two.

1084a 20

But it is absurd also for ten to be a form while eleven is not, nor are the numbers that are next. Also, there both are and come to be some things of which there are no forms, but then why are there not forms of those things? So the forms are not causes. Also it is absurd if number up to ten is more so some sort of being and form than is ten itself, even though there is no coming into being of the former as one thing, while of the latter there is. But they try to make out that number up to ten is something complete. At any rate they generate the derivative things, such as void, proportion, the odd, and other things of that sort within

1084a 30

1078a 32 and footnote), and infinity is always potential in it (see *Physics* Bk. III, Ch. 6). Those who claim that number stands on its own as separate must make it actually infinite or not infinite at all.

the ten numbers; for some things, such as motion and rest, good and bad, they ascribe to the sources, but others to the numbers. (And this is why the one is the odd, for if the odd were in three, how would five be odd?) Also, the magnitudes and the other such things are derived from numbers up to a certain amount, such as first an indivisible line, then two, then those things up to ten.[23]

1084b

Also, if number is separate, one might be at an impasse as to whether the one is prior, or the three and the two. Insofar as a number is composite, the one is prior, but insofar as the universal and the form are prior, the number is; for each of the units is part of the number as its material, but the number is in the manner of form. There is also a sense in which the right angle is prior to the acute, because it is determinate, and by its articulation, but there is another sense in which the acute is prior, because it is a part and the right angle is divided into acute angles. So as material, the acute angle and the element and the unit are prior, but in accordance with the form and the thinghood that is disclosed in an articulation, the right angle and the whole that is made of material and form are prior, since the two together are closer to the form and that of which the articulation is, though later in coming into being. In what way then is the one a source? Because it is not divisible, they say; but the universal is indivisible as well as are the particular and the element, though in a different way, the former in articulation and the latter in time. In which of these ways, then, is the one a source? For just as was said, it seems both that the right angle is prior to the acute and the latter to the former, and each of them is one. So they make the one a source in both senses. But this is impossible, since the one sense is as form and thinghood while the other is as part and material. For each of these is in some way one, for in truth each unit has being as a potency (if, that is, the number is some one thing and not like a heap, but a different number results from different units, as they say), but not as fully at-work.

1084b 10

1084b 20

The cause of the mistake that resulted is that they went hunting at the same time from among mathematical things and from among articulations of universals, and so out of the former they set up that which is one and a source as though it were a point (for the unit

---

[23] The indivisible line refers to the nature of the point (see 992a 22–23), and two is identified with the nature of the line (see 1036b 12–18 and footnote).

is a point without position, so just as some others had put together
beings out of what is smallest, they did too, so that the unit became
the material of numbers, and at the same time prior to two and yet
1084b 30 again derivative from it, since two is a certain whole and is one and
a form). But because they were looking for the universal, they spoke
of the unity that is predicated [of the whole] also in the same way as
a part. But these things cannot belong at the same time to the same
thing.

But if the one itself need only be without position (for it differs in no
way except that it is the first), and two is divisible but the unit is not,
the unit would be more like the one itself, and if the unit is, then the one
itself is more like it than like two, so that each unit in it would be prior
1085a to two. But they deny this; at least they generate the two first. Also if
two itself is one thing and three itself is also, both of them together are
two. But what is this two composed of?

Chapter 9 One might also be at an impasse, since there is no contact
in numbers, but succession of those units which have none in-between
(such as those within two or within three), as to whether they are in
succession with the one itself or not, and whether it is two or one of the
units in it that is prior to what succeeds it. And similarly, inconvenient
results follow also about the kinds of things that are derivative from
number: the line, the surface, and body. Some people make these out of
1085a 10 the forms of the great and the small, lines out of the long and the short,
surfaces out of the wide and the narrow, and bulks out of the deep and
the shallow, and these are forms of great and small. But different people
set down in different ways the source of such things in the one, and in
these, great numbers of impossibilities, fictions, and things contrary to
everything reasonable make their appearance. For the kinds turn out
to be divorced from one another if their sources do not also accompany
one another in such a way that the wide and narrow are also long and
short (but if this is so, a surface would be a line and a solid would be a
1085a 20 surface). Also, how will angles and figures and such be accounted for?
The same thing turns out as with what concerns number, since these
are attributes of magnitude but a magnitude is not made out of them,
for instance length out of straight and curved, or solids out of smooth
and rough.

Common to all these things is the very thing which turns out to
present an impasse in the case of the species of a genus, when one

makes universals stand on their own: whether it is animal itself that is in the animal or something other than animal itself. If the universal were not separate, this would cause no impasse, but if the one and numbers are separate from things, as those who say these things claim, it is not easy to resolve, if one ought to call the impossible "not easy." For when one thinks the unity in the number two or in number in general, does one think some thing itself, or something else?

1085a 30

So some people generate magnitudes out of material of that sort, but others from the point (and for them the point seems not to be the one, but something like the one) and another material that is like multitude but is not multitude,24 about which the same things turn out no less to be impasses. For if the material is one, a line, a surface, and a solid will be the same thing (since out of the same things, one and the same thing will be made), but if the materials are more than one, with one for the line, a different one for the surface, and another for the solid, then either they accompany one another or not, so that the same things follow and in the same way, for a surface will either not have a line in it, or it will be a line.

1085b

Also, they make no attempt to say in what way number can be made out of the one and multitude, but however they mean it, the same inconvenient results follow for them as for those who say it is made out of the one and the indeterminate dyad. For the former generate number out of multitude as universally predicated, and not out of some particular multitude, but the latter from a particular multitude, though the first one (since they say the dyad is the first multitude), so it makes no difference in what way they say it to be, but the same impasses follow, whether it is mixture or placement or blending or becoming or whatever else of that sort. One might especially raise as an impasse, if each unit is one, what it comes from, for each of them is surely not the one itself. So it must be made of the one itself and multitude, or part of multitude. Now to say that a unit is a multitude is impossible, since it is indivisible, but to say it is derived from a part of multitude has many other inconvenient consequences, for each of the

1085b 10

24 In *On the Soul* (409a 4–5), Aristotle refers to people who generate the line by the motion of a point, and then the surface by the motion of a line. This requires a second generating principle, something through which the point can move, which, while not multiplying the point (as the dyad multiplies the one to produce numbers), provides it with a multitude of positions.

But others, since it was necessary, if there were going to be some sort of independent things besides the ones that are perceptible and in flux, that they be separate, since they had nothing else, set up the things that are spoken of universally, so that it followed that the universals and the particulars were just about the same natures. Now this itself, by itself, would be a certain inconvenience for the things said.[27]

1086b 10

Chapter 10 But let us now speak of that which presents a certain impasse for both those who say there are forms and those who say there are not, and was spoken of before among the impasses at the beginning. For if one does not posit that separate independent things have being, and have it in the way that it is attributed to particular beings, one will abolish thinghood in the sense we mean it, but if one does posit separate independent things, in what manner will one set down the elements and sources of them? For if they are particular and not universal, there will be just as many beings as there are elements, and the elements will not be knowable. (For let the syllables in speech be independent things, and let their letters be the elements of independent things; then it is necessary that there be one BA, and one each of the other syllables, if they are not universal and the same in kind but each is numerically one and a this and not something sharing a name—and they do posit that each just-what-it-is is one. But if that is so for the syllables, then it is also that way for the things out of which they are made; therefore there will not be more than one letter A, nor more than one of any of the other letters, by the same argument by which none of the syllables can be the same over and over. But surely if this is so, there would not be any other beings besides the elements, but there would only be elements, and also the elements would not be knowable, since they are not universal, while knowledge is of universals. That is clear from demonstrations and definitions, for no reasoned conclusion comes about that this triangle has angles equal to two right angles unless every triangle has angles equal to two right angles, nor that this human being is an animal unless every human being is an animal.) But surely if the sources are universal, then either the independent things that come from them will also be universal,

1086b 20

1086b 30

1087a

---

[27] Aristotle is having fun here. "Itself by itself" was a common formula for the form understood as a universal, and thus simply duplicating the perceptible world.

or what is not an independent thing will take precedence over an independent thing, for what is universal is not an independent thing, but the element and source is universal, and the element and source takes precedence over the things of which it is the element and source.

Now all these things follow reasonably, when they make the forms out of elements and consider each of them to be some one separate thing apart from the independent things that have the same form. But if nothing prevents it, then just as in the case of the elements of speech there are many A's and B's while there is, apart from the many, no A itself or B itself, there would be, on account of this, an infinity of similar   1087a 10
syllables. But the fact that all knowledge is universal, so that the sources of beings must be universal and not be separate independent things, presents the greatest impasse of those being discussed; however, while what is said is true in a sense, there is a sense in which it is not true. For knowledge, like knowing, has two senses, the one as in potency, the other as at-work. The potency, being, like material, universal and indeterminate, is of what is universal and indeterminate, but the being-at-work is determinate and of something determinate; being a *this* it is of a *this*, but incidentally sight sees a universal color because this color   1087a 20
that it sees is a color, and this A that the grammarian contemplates is an A. Otherwise, if it is necessary for the sources to be universal, it is necessary too that what comes from these be universal, just as in demonstrations, and if this is so, there will not be anything separate nor any independent thing. But it is clear that there is a sense in which knowledge is universal, and there is a sense in which it is not.[28]

---

[28] The main argument of the *Metaphysics*, culminating in Book XII, was that the being of perceptible things depends on forms that are separate and are knowable without being universals. The conclusion here is more limited, that the knowability of perceptible things does not require that there be separate universals. A universal is a general attribute that has to belong to some underlying thing; Book VII showed that the form cannot be conceived as a universal if it is to be a cause. Book XIII does not argue that there are no forms, but that they are not numbers or universals.

Book XIV (Book N)
Intelligible Things as Causes[1]

Chapter 1 Now about this sort of thinghood, let this much have been said. Everyone makes the sources of things contraries, and in the same way as among natural things, similarly too as concerns the thinghood that is motionless. But if it is not possible for anything to be prior to the source of all things, it would also be impossible for the source to be a source while being anything else, as, for example, if someone were to say that white was the source, not insofar as it is anything else but insofar as it is white, and yet that it belonged to some underlying thing and was something else while being white; for then that something else would be prior. But surely all things come from contraries as belonging to an underlying thing; therefore it is necessary that this be present most of all with contraries. Therefore all contraries always belong to an underlying thing and are in no way separate; and in the same way it is apparent that nothing is contrary to an independent thing, and reason bears witness to this. Therefore it is none of the contraries that is the source of all things in the governing sense, but something else.

But these people make one of the contraries material, some making the unequal be material for the one, on the grounds that this is the nature of multitude, others making multitude material for the one (since numbers are generated by some out of the dyad of the unequal, the great and the small, and by another out of multitude, in both accounts by the action of the thinghood of the one). For one who speaks of the unequal and the one as elements, and the unequal as a dyad of the great and the small, speaks of the unequal and of the great and the small as though they were one thing, and does not distinguish that they are one in articulation but not numerically. But they do not even give a good account of the sources which they call elements, some speaking of the great and the small along with the one, these three as the elements of numbers, the two as material and the one as form, others speaking of the many and the few, because the nature of the great and the small is more appropriate to magnitude, still others speaking

1087a 30

1087b

1087b 10

---

1 This title for Book XIV supplied by the translator.

of these in a more general way as what exceeds and what is exceeded. But there is no difference to speak of among these as relates to some of

1087b 20   their consequences, but only as relates to logical hairsplitting, which they guard against because they carry out demonstrations of a logical sort, except that it belongs to the argument itself for what exceeds and what is exceeded, and not the great and the small, to be sources, and for number to come from the elements prior to the dyad, for both of these are more universal. As it is, they say the one thing but not the other. Others oppose the different and the other to the one, while others oppose multitude and the one. But if, as they wish, beings come from contraries, then either there is nothing contrary to the one, or if in the next place something has got to be, then it is multitude, since

1087b 30   the unequal is contrary to the equal, and the different to the same, and the other to something itself, so those opposing the one to multitude most of all have some semblance of a reason—not even they, however, sufficiently; for the one would be few, since multitude is opposed to fewness and the many to the few.

It is clear that the one signifies a measure. And in everything there is something different that underlies it, such as, in harmony, a quarter tone, in magnitude, an inch or a foot or something of that sort, and in rhythm, a step or a syllable, and in heaviness, similarly,

1088a   it is some definite standard weight; and in all things it is the same way, in qualities some quality and in quantities some quantity, and the measure is indivisible, either in its kind or in relation to sense perception, as though there were not some one that is in its own right an independent thing. And this is in accord with reason, since the one signifies that something is a measure of some multitude, and number signifies that a multitude has been measured and is a multitude of measures. (This is why it is also reasonable that the one is not a number, since the measure is not measures, but that which is the measure and the one is a source.) And it is necessary always for something that is the same to underlie everything as a measure, such as a horse as the measure if the things are horses, or a human being if they are human

1088a 10   beings. If the things are a human being, a horse, and a god, the measure is perhaps a living thing, and the number of them will be living things. But if they are a human being, white, and walking, there will be a number of these least of all, because they all belong to something that is one and the same in number, but still the number of them would be of classes or some other such label.

And those who regard the unequal as some one thing, making the dyad an indeterminate mix of great and small, say something very far from anything believable or possible, for these are modifications and attributes of numbers and magnitudes, rather than things that underlie them, the many and few being attributes of number and the great and small of magnitude, just as are even and odd, smooth and rough, and straight and curved. Also, in addition to this mistake, it is necessary that the great and the small and all such things be relative to something, but what is relative is least of all things a certain nature or an independent thing, and it is derivative from the of-what-sort and the how-much; and the relative is a certain attribute of the how-much, as was said, but not material, since there is something else that is material for relations in general, in common, and also for their parts and kinds. For there is nothing great or small, or many or few, or relative in general, that is not some other thing that is many or few or great or small or in a relation.

1088a 20

And a sign that the relative is least an independent thing and a being is that of it alone there is no coming-into-being or passing-away or motion, as there is growth and wasting away with respect to amount, and alteration with respect to the of-what-sort, and local motion with respect to place, and simple coming-to-be and passing-away with respect to thinghood. But there is none with respect to relation, for without being changed, the relative thing will be sometimes greater, sometimes lesser or equal, if the other thing is changed in amount. And the material of each thing must be what is such in potency, and so too with the material of an independent thing, but the relative is not an independent thing either in potency or at-work. It is absurd, then, or rather impossible, to make what is not an independent thing be an element of and prior to an independent thing, for all the other ways of attributing being are derivative from thinghood. Also, elements are not attributes of the things of which they are the elements, but the many and few, both separately and together, are attributed to number, and long and short to a line, and a surface is wide and narrow. But if there is some multitude to which one of them, being few, is always attributed, as in the case of two (for if it were many, then the one would be few), then there should be something that would be simply many, as ten is many if there is nothing greater than this, or ten thousand. In what way then will a number of this sort be made of the few and the many? For then both would need

1088a 30

1088b

1088b 10

to be attributed, or neither, but as it is, one of the two alone is attributed.

Chapter 2 One ought to examine simply the question: is it possible for everlasting things to be composed of elements? For then they would have material, since everything made of elements is composite. If then it is necessary for this material to be that of which it is composed, then even if the thing composed of it always is, if it had come into being it must have come from this, and in every case that which comes to be comes from something that is it in potency (for it could not have come from what neither was nor was potentially it), but what is potential admits both of being-at-work and not being-at-work, then even if as much as possible number, and whatever else has material, always is, it would admit of not being, just like something one day old or however many years old; but if something is of this sort, it could also be something that has lasted so long that there is no limit of it. But then they would not be everlasting, if what admits of not being is not everlasting, according to what followed from other arguments concerned with this.[2] But if what is now being said is true universally, that no independent thing is everlasting unless it is a being-at-work, while elements are material of an independent thing, then in no everlasting independent thing could there be elements as ingredients of which it was composed.

1088b 20

There are some people who make the indeterminate dyad the element that goes along with the one, and make difficulties about the unequal, reasonably, on account of the impossibilities that follow from it; they remove only as many of the inconveniences as follow necessarily for those who speak of the latter on account of making what is unequal and relative an element, but those that are separate from that opinion are necessarily present for them too, whether what they make out of the elements is number composed of forms, or mathematical number. Now there are many reasons for their going astray in connection with these causes, but the greatest of them is that they were at an impasse of an ancient sort, for it seemed to them that all beings would be one thing, being itself, unless one went to battle with and refuted the claim of Parmenides, that "never will this be brought

1088b 30

1089a

---

[2] See 1050b 12–18.

under the yoke, that things that are not *are,*" and that it was necessary
to show that nonbeing is, for in that way, if things are many, beings
would come from the one and something else.[3]

First, though, if being is of more than one sort (for it means, in one
sense, that something is an independent thing, in another that it is
of a certain sort, in another that it is an amount, and the rest of the
ways of attributing being), what sort of one will all things be unless   1089a 10
nonbeing is? Will the independent things be one, or the attributes and
other things likewise, or everything, and will a *this* and a this sort
and a this much and the other things that signify some one thing be
one? But this is absurd, or rather impossible, there having come to
be some one nature that was responsible for what is's being in one
way a *this,* in another an of-this-sort, in another a this-much, and in
another a somewhere. And next, what sort of nonbeing and being do
beings come from? For nonbeing is of more than one sort, since also
being is, and the not-human-being signifies not being a certain *this,*
the not-straight, not being of a certain sort, and the not-three-feet-
long, not being of a certain amount. So from what sort of being and
not-being do beings become many? What is meant is the false, and it   1089a 20
is this nature that is spoken of as the nonbeing from which, along with
being, beings become many, and for this reason it is also said that one
must assume something false, just as geometers assume that what is
not a foot long is a foot long. But it is impossible to be that way, for the
geometers do not assume anything false (since that assumption is not
a premise of the reasoning), nor do beings come from or pass away
into that sort of nonbeing. But since, for one thing, nonbeing is meant
in ranked varieties equal in number to the ways of attributing being,
and besides this, what is meant as false as well as what is meant in
accordance with potency are spoken of as nonbeing, it is from this last
that there is coming-into-being, a human being coming from what is
not a human being but is a human being in potency, and a white thing   1089a 30
from what is not white, but white in potency, similarly whether it is
one thing that comes into being or many.

---

3 This refutation of Parmenides is enacted in Plato's *Sophist,* beginning at 236E. The
connection with the hypothesis of an indeterminate dyad is not explicit there, but
is playfully present in Socrates's original question whether sophist, statesman, and
philosopher are one, two, or three (217A). See Jacob Klein, *Plato's Trilogy* (University
of Chicago Press, 1977), pp. 60–64, the section entitled "Why is 'both' used so often?"

But it appears that the inquiry concerns the manner in which being meant in relation to independent things is many, for the things generated are numbers, lengths, and bodies. But it is absurd to inquire how being in the sense of what things are is many, but not about what sorts or what amounts things are. For surely it is not the indeterminate dyad or the great and the small that are responsible for there being two

1089b   white things, or many colors or flavors or shapes, for then these also would be numbers and units. But if they had gone after these things, they would have also seen the reason why there is manyness in those others, since the reason is the same or analogous. This distortion is also the reason why, in seeking something opposite to being or the one, from which along with them beings come, they posited what is relative, namely the unequal, which is neither contrary nor contradictory to those things, but is one of the natures of beings just like what they are and what sort they are. It was necessary to inquire also into this, in what way things that are relative are many and not one, but as it is,

1089b 10   they inquired in what way there are many units besides the first one, but no longer how there are many unequal things besides the unequal. Yet they use them and speak of great, small, of many, few (from which numbers come), of long, short (from which length comes), of wide, narrow (from which surface comes), of deep, shallow (from which bulk comes), and they speak of still more forms of relative things; so what is responsible with these for their being many?

So it is necessary, as we say, to posit a being-in-potency for each thing. (The one who said these things declared also what it is that is in potency a *this* and an independent thing, but not a being in its own right, namely that it is the relative, as if he had said what is of a certain sort, since it is neither potentially the one or being, nor the negation of

1089b 20   one or being, but is one *of* the kinds of being.) But it was much more necessary, as was said, if one were to inquire how beings are many, not to inquire about the things within the same way of attributing being, how there are many independent things or many sorts of thing, but how *beings* are many, since some are independent things, others attributes, and others relations. So in connection with the other ways of attributing being there is also something else to pause over—how they are many. (For since they do not have being as separate, it is because the underlying thing becomes many things and is of many sorts and amounts; and yet there ought to be some material for each kind, except that this is impossible in separation from independent things.) But in

the case of the *this*, there is some sense in how the *this* is many things,   1089b 30
so long as something is not both a *this* and some sort of nature; the
impasse on that side is rather in what manner there are actively many
independent things and not one. But also, if the *this* and the so-much
are not the same, it is not explained how and for what reason beings
are many, but how they are of many amounts, since every number
indicates some amount, even the unit, because it is a measure and that
which is indivisible in amount. So if an amount and what something is
are different, it is not explained out of what, or in what manner, what   1090a
things are are many; but if they are the same, one still continues to say
many contrary things.

One might also pay attention to the examination, about numbers,
of where one ought to get the belief that they have being. To someone
who posits forms, they offer something to be responsible for beings, if
each of the numbers is a form and the form is for other things somehow
or other the cause of their being (for let this be granted to them); but
to one who does not think that way on account of seeing the inherent
inconveniences that concern the forms, so that it is not for these reasons
that one makes the numbers, but who makes mathematical number,   1090a 10
from what source should he be persuaded that such a number has
being, and why is it useful for anything else? For even the one who
says this[4] does not claim that number is the cause of anything, but
speaks of it as being a certain nature itself by itself, nor does it seem to
be a cause; for all the theorems of arithmetic also apply to perceptible
things, as was said.

Chapter 3 Now those who set it down that there are forms and
that they are numbers, as a result of setting out each of them alongside
the many to get hold of each as some one thing, make an attempt at
any rate to explain in some way why they have being, but since these
things are neither necessary nor possible, one ought not to claim on   1090a 20
this account that number has being. But the Pythagoreans, on account
of seeing many attributes of numbers belonging to perceptible bodies,
made the beings *be* numbers, so they were not separate, but beings were
made out of numbers. And why? Because the attributes of numbers
are present in harmony and in the heaven and in many other things.

---

[4] See 1080b 26–28. The last clause of this paragraph refers to Bk. XIII, Ch. 3.

But for those who say there is only mathematical number, it is not possible by their assumptions to say any such thing, but they said there could not be knowledge of [perceptible bodies]. But we say there is, as we said before. And it is clear that the mathematical things are
1090a 30   not separate, for if they were separate, their attributes would not be present in bodies. So the Pythagoreans are subject to no objection as a result of this sort of thing; however, as a result of making natural bodies out of numbers, things having lightness and weight out of what does not have weight or lightness, they seem to be speaking about another heaven and other bodies, and not about the perceptible ones. But those who make number separate, assume that it has being and is separate because, while the axioms would not apply to perceptible things, the
1090b   things they say are true and appeal to the soul, and similarly with mathematical magnitudes.[5] It is clear, then, that the account opposed to this will say the opposite, and that, for those speaking that way, it is necessary to resolve what was just raised as an impasse—how it is that things that are in no way present in perceptible things can have attributes that are present in perceptible things.

There are some people who, because the point is the limit and extremity of a line, and that of a surface, and this of a solid, suppose that there must be natures of that sort. And one ought to see whether this argument is not too flimsy. For the extremities of things are not
1090b 10   independent things, but they are all instead limits (since even of a walk, and in general of a motion, there is some limit, and it would be a *this* and an independent thing, but that is absurd). And even if they were independent things, they would all belong to these perceptible things (since the argument was stated as applied to these); why then would they be separate?

Also, someone who is not too agreeable might inquire further about all number and mathematical things as to the fact that they contribute nothing to one another as prior things to derivative ones (for if there were no numbers there would nonetheless be magnitudes for those

---

[5] Euclid defines a line as "breadthless length," and all his theorems depend upon its breadthlessness, while perceptibility, even in the imagination, would require breadth. Arithmetic assumes that whatever is countable is subject to the results of adding, subtracting, and so on, but Aristotle points out in the *Physics* (Bk. VII, Ch. 5) that the fact that 100 men can haul a ship 100 yards does not mean that one man can haul it one yard.

who claim that there are only mathematical things, and if there were no mathematical things there would still be the soul and perceptible bodies, but it does not seem from the appearances that nature is a string of unconnected episodes as in a bad tragedy). But for those who posit the forms, there is an escape from this, for they make magnitudes out of material and number, length from the dyad, surfaces perhaps from three and solids from four, or from other numbers, for it makes no difference; but will these things be forms, or what is the manner of their being, and what do they contribute to beings? For these contribute nothing either, just as the mathematical things contribute nothing. And surely there is not even any theorem that applies to them, unless one wants to change mathematics and make up some private doctrines. But it is not difficult, when one has taken up any hypotheses one pleases, to make up and string together a lot of things.

1090b 20

1090b 30

So these people who stick mathematical things onto the forms go wrong in this respect, and those who first made two sorts of numbers, one consisting of forms and the other the mathematical sort, have not said, nor could they say, in what manner and out of what constituents the mathematical sort would have being. For they make it in-between number consisting of forms and multitude consisting of perceptible things. For if it comes from the great and small, it would be the same as that which consists of forms (and if it is one particular sort of small and great, it makes the magnitudes), but if one speaks of something else, he is talking about a larger number of elements; and if the source of each sort of number is a particular one, there will be some one that is common to these, and it will be necessary to inquire in what way the one is these many things, and at the same time how number is generated otherwise than from *the* one and the indeterminate dyad, which is impossible by that account. All these things are irrational and conflict both among themselves and with what is reasonable, and the proverbial tangled web[6] is apparent in them, for it turns into a long story, just as with slaves, when they say nothing sound. And the elements themselves, the great one and the small one, seem to cry out as though they were being pushed around, for they are not at all able

1091a

1091a 10

---

[6] Literally "the 'long story' of Simonides," but the source of the quote is unknown.

to generate any other number than what comes from the one by its being doubled.7

And it is absurd to make up a generation of everlasting things, or rather, it is one of the impossible things. As for whether the Pythagoreans did or did not make up a generation of them, there need be no hesitation, for obviously they say that, when the one had been put together, whether out of planes or surfaces or seeds or out of things they are at a loss to describe, immediately the nearest part of the infinite was drawn in and limited by limit. But since they were generating a world and wanted to speak in a way suited to nature, it

1091a 20   is a just thing to make some examination of them concerning nature, but to let them go from our present pursuit, since we are inquiring about the sources among motionless things, and so must examine the generation of numbers of that sort.

Chapter 4 Of the odd number they say there is no generation, clearly as though there is generation of the even number, and some people construct the even number first from the unequal great and small when they are made equal. So it is necessary that the inequality belonged to them before they were made equal; if they were always equalized, they would not have been unequal beforehand (since there is nothing before what is always the case). Therefore it is clear that they were not making the generation of numbers just for the sake of theoretical analysis.

1091a 30   And there is an impasse, and a censure for those who find it an easy passage, as to how the elements and sources are related to the good and the beautiful; the impasse is this, whether one of the sources is the sort of thing we mean by speaking of the good itself and the highest good, or this is not one of them but something generated as derivative from them. Things that come down from those who wrote about the gods seem to agree with some people of the present time who say that the good and the beautiful are not sources but make their appearance

---

7 The sentence begins by making fun of the attempt to use the same elements to make two kinds of numbers, but Aristotle cannot resist tossing in a participle at the end to remind us that those elements cannot even account for all the numbers in one series, since the dyad applied to the one will produce only the series of doubles 2, 4, 8, etc.

within the nature of things when it has advanced. (They do this out
of caution about a true difficulty which follows for those who say, as
some do, that the one is a source. The difficulty is not on account of
reckoning what is good to the source as something present in it, but on
account of making the one a source—and a source in the sense of an
element—and making number out of the one.) And the ancient poets
speak in a way similar to this, insofar as they say that it is not the
first gods, such as Night and Heaven, or Chaos and Ocean, that reign
and rule, but Zeus. But it turns out that these poets say such things
by way of saying that the rulers of beings change, although the ones
who are of a mixed sort by not saying everything in myths, such as
Pherecydes and some others, set down the highest good as the first
begetter, as do the Magi and, among the later wise men, for instance,
Empedocles and Anaxagoras, the former making love an element and
the latter making intelligence a source. Of those who say the thinghood
of things is motionless, some say that the one itself is the good itself;
however, they believed its thinghood to be most of all its oneness.

1091b

1091b 10

The impasse, then, is this: in which of the two ways ought one to
speak? But it would be surprising if, in what is primary and everlasting
and self-sufficient, this very self-sufficiency and self-maintenance were
not present primarily as good. But it is certainly not on account of
anything other than that it is in a good condition that it is indestructible
or self-sufficient; and so, it is reasonable that it is true to claim that the
source of things is of this sort, but for this to be the one, or if not that at
least an element, and an element of numbers, is impossible. For many
inconvenient consequences follow, to escape from which some people
have renounced the claim, who agree that the one is a first source and
element, but only of mathematical number; for all the units would
become things the very nature of which is goodness, and there would
be a vast abundance of goods. Also, if the forms are numbers, the very
nature of every form would be goodness; but then let someone posit
that there are forms of whatever he wants—if there are forms only of
kinds of good, there will be no forms that are kinds of thinghood, or
if there are also forms of the kinds of thinghood, all the animals and
plants and everything that participates in a form will be good.

1091b 20

1091b 30

These absurd consequences follow, and it also follows that the
opposite element, whether it is multitude or what is unequal, the
great and small, is the bad itself. (For this very reason one person

avoided attaching goodness to the one on the grounds that it would be necessary, since coming into being is from contraries, that the nature of multitude would be badness; others do say that inequality is the nature of the bad.) So it follows that all beings partake of badness—outside of one, the one itself—and that numbers partake of a more unmixed badness than do magnitudes, and that the bad is the place of the good, and partakes in and stretches out toward that which tends to destroy it, for a contrary is destructive of its contrary. And if, as we said, material is each thing in potency, as the material of fire at-work is fire in potency, then the bad itself would be the good in potency. All these consequences follow, for one reason because they make every source an element, for another because they make contraries sources, for another because they make the one a source, for another because they make numbers the primary independent things, and make them separate, and make them forms.

1092a

Chapter 5 Now if it is impossible *not* to set down the good among the sources, and also impossible *to* set it down in this way, it is clear that the sources are not being accounted for rightly, and neither are the primary independent things. Nor does one conceive them rightly if one likens the sources of the whole to that of animals and plants, in that what is more complete always comes from what is indeterminate and incomplete, on which account one says it is that way even with the primary things, so that the one itself is not even a being. For even here the sources from which things come are complete, for a human being begets a human being, and it is not the seed that is first. And it is absurd to make place be together with mathematical solids (for the place of particular things is peculiar to them, which is why they are separate in place, while mathematical things are not anywhere), and also absurd to say that they are somewhere but not say what that place is.

1092a 10

1092a 20

Those who say that beings come from elements and that the first of the beings are numbers ought to have distinguished in what ways one thing comes from another, and thus explained in what manner number comes from the sources. Is it by a mixing of them? But not everything can be mixed, and what comes from it is something different, so the one would not be a separate thing or a distinct nature, which they want it to be. Is it then by placing them together as in a syllable? But then they would need to have position, and also, one who thinks of the one and multitude would think them as separate. This, then, would be number: a unit plus multitude, or the one plus the unequal. And since there is a sense in which coming from things means from things present as constituents, and a sense in which it does not, which way is it with number? For the way in which it is from things present as constituents is nothing other than the way there is generation out of something. But then does number come about as from a seed? But it is impossible for anything to have come out of what is indivisible. Then does it come as from a contrary that is not still present in it? But things that come about in this way also come from something else that is still present in them. Since, then, one person sets down the one as contrary to multitude, another as contrary to the unequal, treating the one as the equal, number would come about as from contraries, and therefore there is some other thing still present out of which, along with one of the contraries, number is made or has come to be. Also, why in the world do all other things that come from contraries or have contraries perish, even when they are made from all of it, but not number? Nothing is said about this. Yet whether it is present in it or not present in it, the contrary destroys a thing, as strife destroys what is mixed (and yet it shouldn't, since it is not that to which it is contrary).[8]

1092a 30

1092b

But also, nothing has been distinguished about the way in which numbers are causes of independent things and of being; is it as limits (as points are limits of magnitudes, and as Eurytus assigned what was the number of what, such as this number for a human being and that number for a horse, in the same manner as those who arrange numbers into the shapes of a triangle or quadrilateral, thus imitating

1092b 10

---

[8] Aristotle's analysis of becoming in I, 7, of the *Physics* discovers three sources of anything that comes to be: the form which it comes to have, the contrary or deprivation of that form, out of which it changed, and the underlying material in which the form is present. Whether the dyad is understood as contrary to the one, or as material for it, something is missing in this two-source hypothesis. In any event, this account of coming into being applies only to things capable of change, in which a tension of contraries is always present. It makes no sense for anything changeless and everlasting to have a genesis. The last example, and the irresistible criticism of it, refer to Empedocles.

with pebbles the shapes of natural things⁹), or is it because harmony is a ratio of numbers, and likewise a human being and each of the other things is too? But how are the attributes, such as white, sweet, or hot, numbers? That numbers are not thinghood nor causes of form is clear, for the ratio is the thinghood but the number is the material. For example, a number is the thinghood of flesh or bone in this way:

1092b 20 three parts fire and two of earth. The number, whatever it may be, is always a number of something, either of fire or of earth or of arithmetic units, while the thinghood is there being so much to so much in the mixture, and this is no longer number but a ratio, in a mixture, of numbers of bodies or of whatever else. So number, either number in general or number consisting of arithmetic units, is not a cause either as producing anything, or as material or articulation or form of things, and it is certainly not a cause in the sense of that for the sake of which.

Chapter 6 One might also be at an impasse as to what the good is that comes from the numbers by a mixture's being in number, whether in an easy ratio or in an unusual one. As it is, honey-water is no more healthy when it is mixed three-to-three, but it would do more good if

1092b 30 it were in no particular ratio but watery, than if it were in a numerical ratio but not very diluted. Also, ratios of mixtures are meant in terms of addition of numbers, not in numbers of numbers, that is, as three-to-two and not three times two. For the same kind of thing has to be present in multiplications, so that the product ABC must be measured by A, and the product DEF by D, and all products by something that

1093a is the same. Therefore the number of fire could not be BECF and that of water BC.¹⁰

---

⁹ Geometrical shapes have an inherent affinity to various series of numbers. We identify some numbers as squares or cubes. This is worked out more extensively in the Pythagorean tradition, and may be seen in Nicomachus of Gerasa's *Introduction to Arithmetic*, Bk. II, Ch. 8–12. The extension of it to natural shapes is low comedy. At 993a 14–25 Aristotle makes fun of the assignment of ratios to the thinghood of things, as a confused and childish attempt to understand form as a cause.

¹⁰ Aristotle writes, for the number of water, two times three, rendered here by the letters of the alphabet that would be the corresponding numerals. He could mean that fire would then be just a larger quantity of water (if E and F are just multipliers), or perhaps that water could not be determined by a bare product of numbers while fire was a mixture of materials (E and F) in that ratio.

But if it is necessary for all things to partake of number, it must follow that many things are the same, and that the same number belongs to this thing and to that other. So then is this the cause and is it on account of this that the thing *is*, or is this unclear? For example, there is some number of the revolutions of the sun, and in turn of those of the moon, that is also the number of the life and the age of each of the animals. So what prevents some of these from being squares and some cubes, some equal and some double? Nothing prevents it, but it is necessary that they turn up among these, if all things were to share in number, and different things were to be capable of falling under the   1093a 10 same number, so that if there turned out to be the same number for some of them, they would be the same as one another, since they have the same form with respect to number; for instance, the sun and moon would be the same. But why are these causes? There are seven vowels, a musical scale has seven strings, the Pleiades are seven, at seven the teeth fall out (for some animals, but some not), and there were seven against Thebes. Is it, then, because the number is of a certain sort by nature, that for this reason the attackers turned out to be seven or the Pleiades to consist of seven stars? Or were the former seven on account of the gates, or for some other reason, while the latter we count that way, as we count twelve stars in the Bear, while others count more.

Indeed, people even say that the letters ksi, psi, and dzeta are   1093a 20 concordant intervals, and because there are three of those, these also are three, but they pay no attention to the fact that there could be thousands of such double letters (since there could be a single sign for g and r); but if they say each of these is a double of other letters, and no other letter is, and the reason is that there are three places[11] in each of which one letter is imposed upon the sigma, then this is why there are only three, and not because there are three concordant intervals, since in fact there are more than three concordant intervals, but in the other case it is no longer possible. These people are like the ancient Homeric interpreters who saw small correspondences but overlooked large ones. And some people say that there are many things of this sort, for example the intermediate strings in a scale have the lengths nine and eight, and a line of verse has seventeen syllables, equal in number   1093a 30 to these, scanned in one half-line by nine syllables and in the other by   1093b

---

[11] The palate, lips, and teeth, for ksi, psi, and dzeta, respectively.

eight.[12] They say also that the interval in letters from alpha to omega is equal to that from the lowest to the highest note on flutes, the number of which is equal to the whole tuning of the cosmos.[13] One ought to see that no one would be at a loss to state or to discover such things among everlasting things, since they are present even in destructible things.

1093b 10   But the renowned natures present in numbers, and the contraries of them, and generally what is present in mathematical things, that some people speak of and make into causes of nature, seem to those who examine them in this way to vanish (for in none of the ways that have been distinguished regarding sources of things—in *none* of these— are they causes). It is the case, though, as they make clear, that what is well made is present, and that the odd, the straight, the equal-times-equal, and the potentialities of some numbers belong in the list that includes the beautiful; for seasons go together with a certain sort of number, and all the other things that they collect from mathematical theorems have this potency. For this reason it is fitting that they have correlations,[14] since they are incidental attributes, but all with affinities to one another, and they are one by analogy. For in each designation

1093b 20   of being there is something analogous: as the straight is in length, so is the plane in a flat surface, and perhaps the odd in number and the white in color. Also, it is not the numbers among the forms that are the causes of harmonies and such things (for equal numbers of that sort

---

[12] The intermediate strings are for the fifth (2:3) and the fourth (3:4); in whole numbers, the fundamental string would be six units long, the fifth nine, and the fourth eight. Homer's dactylic hexameter line, with a trochee in the last foot, has seventeen syllables; with a light pause (caesura) in the middle, a normal line would break into nine and eight syllables.

[13] There are 24 letters in the Greek alphabet and 24 notes on a flute; there are various conjectures about how to find the same number in the heavens. Playful examples of the "tuning of the cosmos" may be found in Plato's *Timaeus,* 35A–36B, and in Kepler's seventeenth century *Harmonies of the World.*

[14] At the beginning of the section called the Third Day of his 1638 work *Two New Sciences,* Galileo uses this same Greek word (*symptomata*), in the midst of a passage in Latin, to describe the topics of his new science of motion, including the parabolic path of projectiles and the fact that falling bodies cover distances that are to one another as the squares of their times. He agrees with Aristotle that they are not causes of the phenomena, but things that invariably accompany them. By focusing attention on them to the exclusion of perceptible bodies, Galileo effectively initiates the mathematical physics that has claimed for itself the name science.

differ in kind, since even their units are different), so that, on account
of these things at any rate, one need not make there be forms.

These, then, are some of the things that follow, and still more could
be brought together. But the many respects in which they fare badly
in regard to the generation of them, and their inability to make them
coherent in any way, seem to be a sign that mathematical things are
not separate from perceptible things, as some people say, and that they
are not sources of things.

Bibliography of Suggested Reading

About Aristotle's thinking in general

Jacob Klein, "Aristotle, an Introduction," in *Ancients and Moderns*, edited by Joseph Cropsey (Basic Books, New York, 1964), and in Klein's *Lectures and Essays* (St. John's College Press, Annapolis, Md., 1985). The best short introduction to Aristotle available.

Henry Veatch, *Aristotle, a Contemporary Appreciation* (Indiana University Press, 1974). A good, unpretentious short book, with an emphasis on Aristotle's logic.

Marjorie Grene, *A Portrait of Aristotle* (University of Chicago Press, 1963). More substantial but very readable, with an emphasis on Aristotle's biology.

About the *Metaphysics*

Charlotte Witt, *Substance and Essence in Aristotle* (Cornell University Press, 1989). Brief and clear.

Mary Louise Gill, *Aristotle on Substance, the Paradox of Unity* (Princeton University Press, 1989). More detailed and technical, with an emphasis on the meaning of material.

Edward Halper, *One and Many in Aristotle's Metaphysics, the Central Books* (Ohio State University Press, 1989). Still more technical; a very faithful account of the central argument, in all its complexity.

Joseph Owens, *The Doctrine of Being in the Aristotelian Metaphysics* (Pontifical Institute of Mediæval Studies, Toronto, 1951). A massive work of scholarship. Head and shoulders above any other commentary on Aristotle produced in our time.

Martin Heidegger, *Aristotle's Metaphysics Θ 1-3* (Indiana University Press, 1995). Fresh, penetrating, and suggestive.

About other works of Aristotle

Martin Heidegger, *Plato's Sophist* (Indiana University Press, 1997). The first part, about a third of the book, is a study of Aristotle's treatment of the intellectual virtues in the *Nicomachean Ethics*, with numerous connections to his other works. Groundbreaking work, from one of Heidegger's earliest lecture courses.

Yves Simon, *The Great Dialogue of Nature and Space* (Magi Books, 1970). A transcribed course of lectures dealing primarily with the *Physics*; an excellent introduction.

Martha Nussbaum and Amélie Rorty (eds.), *Essays on Aristotle's De Anima* (Clarendon Press, Oxford, 1992). Unusually lively and focused.

Index

This is not a complete index to the *Metaphysics*, but a help in finding one's way to some of its main references to topics and persons, as well as to those in the Introduction and footnotes. The outline of Aristotle's argument and the Glossary, following the Introduction, should also be consulted. More comprehensive indexes, keyed to the Greek text, may be found at the back of W. D. Ross's two-volume text and commentary, *Aristotle's Metaphysics* (Oxford U. P., London, 1958), and in Hermann Bonitz's *Index Aristotelicus* (De Gruyter, Berlin, 1961).